LABORATORY EXERCISES & TECHNIQUES IN CELLULAR BIOLOGY

LABORATORY EXERCISES & TECHNIQUES IN CELLULAR BIOLOGY

Anthony Contento
State University of New York at Oswego

WILEY

VP & EXECUTIVE PUBLISHER	Kaye Pace
ACQUISITIONS EDITOR	Kevin Witt
ASSISTANT CONTENT EDITOR	Lauren Morris
MARKETING MANAGER	Clay Stone
DESIGNER	Kenji Ngieng
ASSOCIATE PRODUCTION MANAGER	Joyce Poh

Cover image: Micrograph showing the generation and differentiation of iPSCs (Courtesy Fred H. Gage and Kristen Brennand).

This book was set in 10.5/13 Times Roman by MPS Ltd, Chennai. Cover and text printed and bound by Bind Rite.

This book is printed on acid free paper. ♾

Founded in 1807, John Wiley & Sons, Inc. has been a valued source of knowledge and understanding for more than 200 years, helping people around the world meet their needs and fulfill their aspirations. Our company is built on a foundation of principles that include responsibility to the communities we serve and where we live and work. In 2008, we launched a Corporate Citizenship Initiative, a global effort to address the environmental, social, economic, and ethical challenges we face in our business. Among the issues we are addressing are carbon impact, paper specifications and procurement, ethical conduct within our business and among our vendors, and community and charitable support. For more information, please visit our website: www.wiley.com/go/citizenship.

Printed in the United States of America

10 9 8 7 6 5 4 3 2 1

CONTENTS

Author's note: Exercises in the Table of Contents marked with an * are designed as online content. The experiments discussed in these exercises are either "dry," requiring only student discussion and Internet research; **or** they are advanced experiments, requiring specialized reagents or equipment that might not be found in all teaching laboratories. The remaining twenty exercises will be available in the printed manual. Access to all exercises will be available to students at the time of purchase of the printed manual.

INTRODUCTION

Applied Cell and Molecular Biology Laboratory Manual engages students by focusing on the physiology, structure, and function of eukaryotic cells. The lab manual itself consists of 20 experiments that coincide and complement each of the 18 chapters in Gerald Karp's *Molecular Cell Biology: Concepts and Experiments*. Additional experiments covering advanced topics will also be provided online, freely accessible with the purchase of the manual on the book companion site (www.wiley.com/college/contento). Illustrations and photographs are used throughout the book to emphasize proper technique and illustrate expected results.

The exercises within each lab are divided into five sections.

- An **Overview** section can be found at the beginning of each lab. *Objectives* and *Description* sections outline the specific goals of each lab and explain the importance of each technique. *Concepts & Vocabulary* details the appropriate terms specific to each lab topic. The applications of common laboratory techniques in science, medicine, and everyday life are also explored in each lab topic in a topical *Current Applications* section. *References* back to Karp, as well as to important papers in the scientific literature are provided at the bottom of every **Overview**.
- A strong **Background section** is provided with each experiment. This expanded introduction resonates with information found in Karp, creating synergy between the textbook and the lab manual. The Background section is also robust enough to provide enough information to use this manual for use with another textbook or as a stand-alone laboratory manual.
- **Step-by-step Procedures** are provided so that students can successfully complete each experiment in a timely manner. These procedures are connected to a detailed Instructor's Guide which allows for proper preparation and planning of each exercise.
- **Engaging Worksheets** are designed to encourage proper data collection, as well as challenge the students with careful analysis of their results.
- **Discussion** questions are provided to initiate in-class or online dialogue about the important concepts and applications surrounding each topic/technique. Through these conversations, more advanced material may be introduced in response to specific student questions.

In addition to the overall approach and design, this laboratory manual:

- **Provides a clear and complete introduction** to a number of common biochemical, physiological and histological techniques used in contemporary Molecular Cell Biology. These experiments should provide students with a solid example of the technique, and offer hands-on experience with proper application. Application of each technique should offer verification of the concepts introduced in the Karp text, and allow them to add another item into the "toolbox" of protocols that science students should be amassing during their college studies. Group and cohort paths to knowledge are encouraged in order to divide resources and time between several students.
- **Allows for all possible learning techniques**, including as verification of key concepts, technique introduction, inquiry-based learning, and research-based active learning experiences. The techniques may be taught as they are and provide excellent verification of concepts in the text, or they may be used as introductory exercises for an inquiry-based or active-learning research

project. Students may be asked to design their own projects; engage in directed, on-campus, undergraduate research; or they may become engaged in a learning program that attempts to provide research experiences for undergraduates. The CASPiE (Center for Authentic Science Practice in Education, http://www.purdue.edu/discoverypark/caspie/) project is designed to create a "real-world," active-learning research experience. It uses specific techniques-driven introductory projects that lead to an actual research project designed to cater to the input of student data.

- **Develops a variety of additional techniques** that include traditional laboratory activities, Internet research, calculations/extrapolations, and critical analysis. Because the pursuit of real-world cell biology involves all these components, so do our lab activities.
- **Emphasizes the improvement of written and other forms of scientific communication**. So much of science has become participatory, particularly in making decisions about its application. We provide ways for the student to discover that the communication of scientific information is as important as the acquisition of scientific knowledge.
- **Contains relevant problem sets** that can be used as labs, lab supplements, or homework assignments for cell biology lectures.
- **Contains relevant discussion questions** that can be used in lab discussions, supplemental homework assignments for cell biology lab and/or lectures. In combination with moral conversation techniques, a number of difficult dialogues important to cell biology and the biological sciences may be broached by the instructors and students.

Applied Cell and Molecular Biology Lab Manual can be used with any Cell and Molecular Biology text, but is available for custom packaging with Karp's *Cell and Molecular Biology* by visiting customselect.wiley.com or by contacting your Wiley Representative.

ACKNOWLEDGEMENTS

The author would like to thank Dr. Martin Spalding and the Iowa State University Department of Genetics, Development and Cell Biology for their support and the use of their teaching laboratory facilities for the development and testing of the laboratory exercises found in this manual.

The author would also like to thank Dr. Diane Bassham and Dr. Steven Rodermel, as well as Iowa State University Genetics Laboratory Coordinator Claudia Lemper, for their suggestions and advice during the planning for this book.

Artist: Grant Robbins
Student Sketches: Erinn Rieser
Photographs and Micrographs: Anthony Contento
Student Experimental Testers: Melissa Berryman, Mone Contento, Angela French, Valerie Giles, Heidi Horstman, Joshua Kurtenbach, Mckenzie Lambert, Audra Lloyd, Erinn Rieser, Yasaira Rodriguez, Lavanya Singh

The author would also like to thank those individuals that reviewed the lab manual prior to publication:

Sylvie Bardin, University of Ontario, Institute of Technology
Blake Bextine, University of Texas-Tyler
Ashok Bidwai, West Virginia University
Gail Breen, University of Texas-Dallas
Xio Wen Cheng, Miami University-Oxford
Peter Chung, Pittsburgh State University
Elizabeth Dudkin, Penn State University-Brandwine
Pamela Elf, University of Minnesota-Crookston
Harold Hoops, SUNY Geneseo
Rebecca Kellum, University of Kentucky-Lexington
Margaret Kenna, Lehigh University
Cran Lucas, Louisiana State University- Shreveport
Anne Luebke, University of Rochester
Thomas McGuire, Penn State-Abington
Peter Meberg, University of North Dakota
James E. Raynor, Jr., Fayetteville State University
Montserrat Rebago Smith, Kettering University
Susan Rouse, Southern Wesleyan University
Laura Schramm, St. John's University
David Schultz, University of Louisville
Joel Sheffield, Temple University
Jia Shi, University of Colorado
Richard Showman, University of South Carolina
Sheldon Steiner, University of Kentucky- Lexington
John Sternfeld, State University of New York-Cortland
Jose Vazquez, New York University
Carey Waldburger, William Patterson University
Andre Walther, Cedar Crest College
Dara Wegman-Geedey, Augstana College
Michael Yeager, MCSD

OVERVIEW 1

Introduction and Microscopy

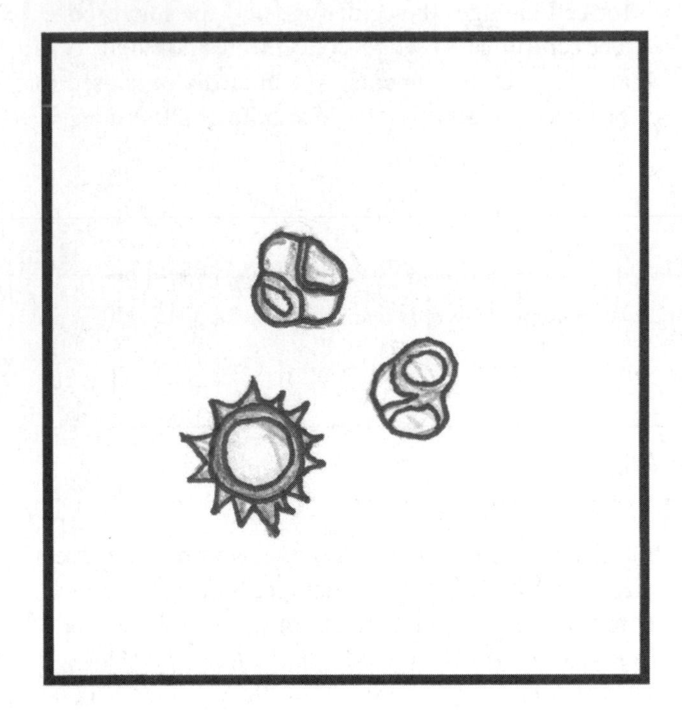

Objectives

- Discuss the three possible types of experiments in this manual.
- Perform a simple centrifugation.
- Review the proper use in the setup of the light microscope.
- Work with dry and oil objectives.
- Calibrate an ocular scale using a micrometer.
- Observe several different types of slide preparations.

DESCRIPTION

Cell Biology is concerned with study of the morphology, physiology, and development of the cell. This manual will present educational experiments that touch on each of these three topics. Histology experiments deal with the preparation of cells or tissues for microscopic observation. Wet laboratory experiments present applied biochemical protocols that can be used to observe cellular physiology and other processes. Dry laboratory experiments do not involve microscopes, or chemicals and test tubes, but instead involve the use of computers, software, and online tools, as well as Socratic exchange, to discuss the application of broader concepts that cannot easily fit into a three-hour laboratory session. In this exercise we will review the use of equipment commonly used in molecular cell biology. We will also go over safe laboratory practices. Finally, we will review proper tuning and usage of the microscope, as you will be using it often to complete the exercises in this manual. We will discuss the different types of microscope and illumination, and you will calibrate a microscope and use it to observe several prepared and fresh slides.

CONCEPTS & VOCABULARY

- Centrifuge
- Condenser
- Histology
- Light microscopy
- Micrometer
- Numerical aperture
- Objective
- Ocular
- PPE
- Resolution

CURRENT APPLICATIONS

- Cell biology would not be a unique discipline with microscopy and histology. The ability to stain and observe the parts of the cell in fine detail has led to numerous discoveries in biology and medicine. Today, there are many kinds of microscopes with different types of optics, illumination, and imaging systems.
- There are two pieces of equipment that are essential for cell biology: the centrifuge and the microscope. The centrifuge is used to concentrate and isolate cells and cellular components. The microscope is used to observe tissues, cells, and intracellular structures.

REFERENCE

Karp, G. (2010). Microscopy and centrifugation. *Cell and Molecular Biology: Concepts and Experiments.* (715–730; 733). Hoboken, NJ: John Wiley & Sons, Inc.

BACKGROUND

APPLIED MOLECULAR CELL BIOLOGY

Cell biology is an integrated discipline of the biological sciences that is focused on the nature of the cell. Cell biology includes everything from cellular physiology to cell morphology and cell interactions with the environment. As an integrated discipline, cell biology borrows techniques from the other biological disciplines. Some experiments involve biochemical processes, while others involve simple observational analysis. The discipline can focus on the intracellular and extracellular events of a single cell, a specific tissue, an organ, or an entire organism. Many cell biologists focus their entire research on a single organelle or a single intracellular physiological process.

Applied molecular cell biology involves the practice of the techniques unique to the discipline; while borrowing from biochemistry and molecular biology, most microscopy, cell culture, and cell fractionation techniques are unique to cell biology. This manual has been designed to introduce a variety of common techniques used by the modern cell biologist. While all of these experiments cannot be covered in a single university term or semester, each experiment has been designed in an instructive manner to engender familiarity with the technique.

Each experiment contains five sections. The Overview sections highlight the basic goals of the technique, common jargon and applications, and relevant passages in Gerald Karp's *Cell and Molecular Biology: Concepts and Experiments*. The Background section provides a brief introduction to the technique and the science behind it. The Procedures section offers step-by-step directions for each experimental technique. The Discussion section provides thoughtful and thought-provoking questions that can be discussed with your peers and your instructors in order to increase your understanding of the experiment. Finally the Worksheet section provides a space for observational records and data analysis. Each experiment also includes illustrations and photographs of relevant techniques, equipment, and examples of data. There is also additional online content available for students interested in advance techniques.

TYPES OF EXPERIMENTS

This manual contains three types of experiments: wet, histology, and dry experiments. Wet experiments are the student laboratory exercises that most of you have experienced in other classes. They involve biochemical analysis, separation of cellular components, and molecular biology techniques. Histology experiments involve the staining or labeling of molecules or structures inside or near the cell followed by microscopic observation. Dry experiments involve the use of digital or online resources or demonstrations and discussions of advanced techniques and cell biology theory. Together, these three types of experiments encompass most of the techniques used in contemporary applied molecular cell biology.

SAFETY AND PERSONAL PROTECTIVE EQUIPMENT

There are a number of chemical and infectious hazards associated with cell biology experimentation. Foreknowledge of the risks and preparation of proper equipment are the best ways to reduce the risk of accidents and injuries. Barrier protection should be used at all times. Barrier protection includes UV protective safety goggles, non-latex examination gloves, and a lab coat. Additional precautions include wearing closed-toe shoes, tying back long hair, and removing any unnecessary objects from your lab bench. Be sure to clean your work area and wash your hands before leaving the lab. Never use consumables (food, beverage, make-up, etc.) in the lab. If a spill or an accident occurs, report it to an instructor immediately. Pay attention to the instructor for details on how to properly dispose of certain chemicals. There are specialized containers for sharp items, such as broken glass or cutting blades and other biohazard wastes. Specialized pieces of equipment, such as a fume hood, will be required when using certain chemicals. There may also be safety guidelines specific to unique usage of equipment (centrifuge and microscopes that you will be using). Always familiarize yourself with these safety guidelines prior to coming to class. Do not hesitate to ask questions if you are unsure as to how to proceed safely during an experiment.

CENTRIFUGATION

Centrifugation is one of the essential techniques in cell biology. Centrifugation allows us to separate a sample based on the density of the components held in a liquid suspension. Centrifugation uses a device called a centrifuge, which consists of a rotor attached to a motor, which causes rotational movement. Tubes containing liquid suspension that are placed into a centrifuge rotor will be affected by centrifugal force. Centrifugal force will pull the denser portions of the liquid suspension to the bottom of the tube. This allows the cell biologist to separate the components of the solution by density. Today you will perform a simple low-speed centrifugation to familiarize yourself with the equipment. In the experiment that follows, you will use other types of centrifuges and other types of centrifugation techniques.

LIGHT MICROSCOPY

The first microscope was invented over four hundred years ago. Since that time microscopy has become more and more complex allowing greater magnification of objects invisible to the naked eye. Aside from the centrifuge, the light microscope is an essential piece of equipment for a cell biologist. The light microscope uses a white light source of varying intensity combined with at least two sets of magnifying lenses in order to offer magnifications of anywhere from 20× to 1000×.

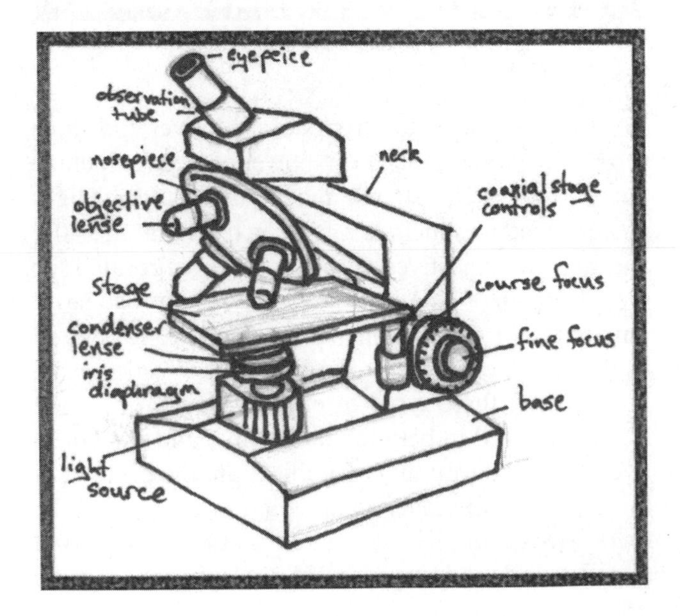

The light compound microscope is composed of several parts. The light source may be a mirror reflecting ambient light toward the objectives or may be an electric light source. The intensity and the focus of the light source are controlled by the iris diaphragm and condenser lens, respectively. A sample that is being observed under the microscope will sit on the stage, usually on top of a glass slide or a clear plastic culture plate. The position of the sample on the stage is controlled by coaxial stage controls. The position of the stage itself in reference to the optics is controlled by the coarse and fine-focus knobs. The optics includes the eye-piece or ocular objective, which is the part that you look through. Beyond the ocular is a series of reflective mirrors that focus the sample image through an objective lens. There may be a single objective lens on a microscope, but typically there is an objective wheel with several different lenses of varying magnifications.

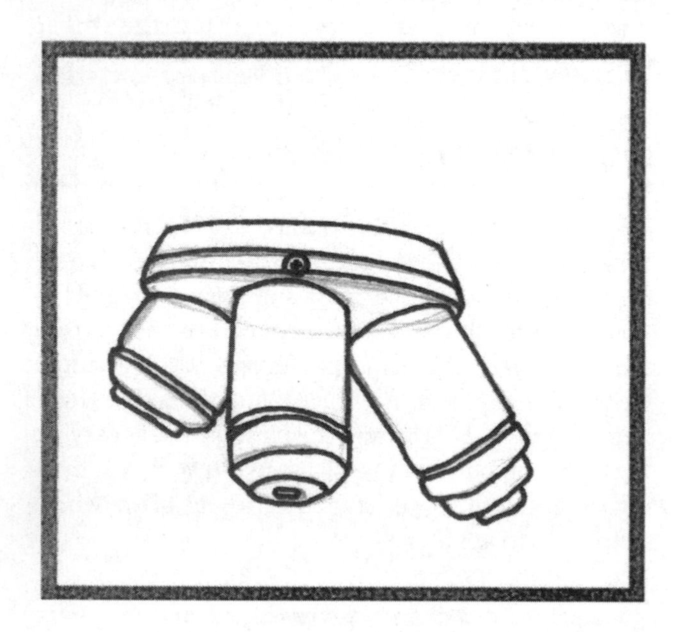

There are two main types of objectives and three sub-types of objectives. Low-power objectives include any objectives of 20× magnification or lower. High-power objectives include any objectives that are 40× and higher. Objectives may be dry, or calibrated for the refraction of light in air. Objectives may also be oil immersion or water immersion, which are calibrated to the refracted indexes of oil or water, respectively. When using these types of objectives, a drop of the liquid is placed in between the sample and the objective. When determining magnification, the product of the objective and the ocular magnification must be used. Most oculars have a magnification around 10× (you can find the exact magnifications engraved on the ocular itself). Most objectives have magnifications between 5× to 100× (also engraved on the ocular).

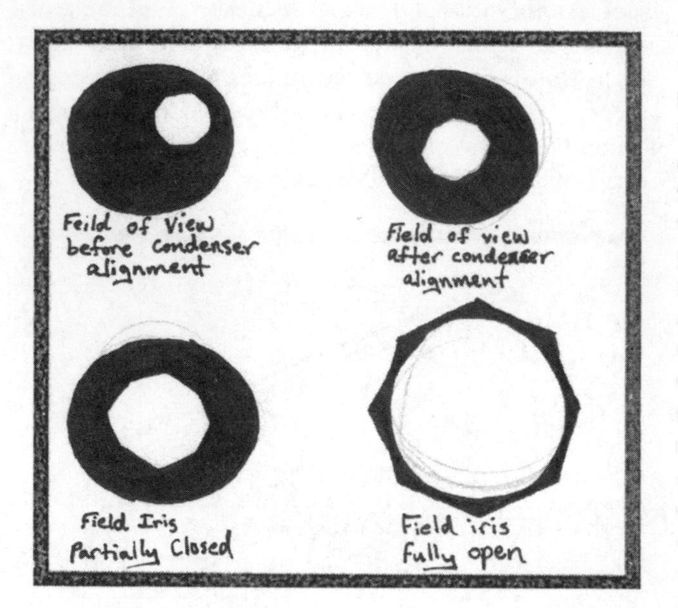

In order to obtain uniform illumination, a microscope must be adjusted for Kohler illumination. This is achieved by focusing the illumination at the plane where the condenser iris can be observed and then focusing an image of the iris diaphragm at the plane of the object you wish to observe. After you have focused your illumination, you will adjust the condenser lens to offer the maximum focused illumination.

HISTOLOGY

Histology is the preparation of samples for microscopic observation. This includes everything from simply putting a sample onto a microscope slide and adding a cover slip to staining specific elements of the sample, or even embedding the sample in paraffin wax and creating thin sections of a tissue or cell type. Most histological samples must be suspended in liquid or dried onto a microscope slide. A cover slip prevents sample contamination of the objectives and vice versa. There is a wide variety of techniques that can chemically alter a sample using redox chemistry, organic changes, or various chromophores. To highlight specific molecules within a cell, histological samples can be fixed using various chemical preservatives to maintain structure and avoid enzymatic breakdown, or they can be prepared as live cells, also known as a vital mount. Histological labeling can be observed using simple light microscopy, but there are some labels that use other forms of illumination.

OTHER FORMS OF MICROSCOPY

With the standard light microscope, a biologist can observe a sample with a magnification of up to 1300×. Using UV fluorescent microscopy and fluorescent labels, greater resolution of intercellular components can be achieved due to the intense signals of fluorescent chromophores. There are also microscopes that use other sources of sample illumination. The laser confocal microscope allows for specific excitation of fluorescent labels and even greater resolution of intercellular components. The transition electron microscope uses an electron beam as an illumination source and can give access to magnification up to 1000× greater than light microscopy. However these more powerful kinds of microscopy require special kinds of preparation. Competence and familiarity of the microscopes in the laboratory will serve you well in the semester. If you find that you enjoy microscopy, ask your instructor about other opportunities on campus. It is an excellent skill to add to your toolbox as a biologist.

MICROSCOPE SAFETY

- Always be aware of where your objective is located.
- It is very easy to break a slide or objective when focusing on a sample sitting on a stage.
- A good rule of thumb is to use the rough focus knob only when you are not looking through the ocular.
- You can use the fine focus knob when observing through the ocular, but be careful to never let the objective touch the cover slip.
- Be sure to remove any oil or sample from the objective as soon as you are finished with using the microscope; only use lens paper and non-ammonia glass cleaner to clean your objective.
- Your coarse and/or fine objective knob may have a locking mechanism in the form of a bar or button. Always make sure the lock is disengaged before trying to move the stage. If you fail to do so, you can strip the gears that control focusing and stage movement.

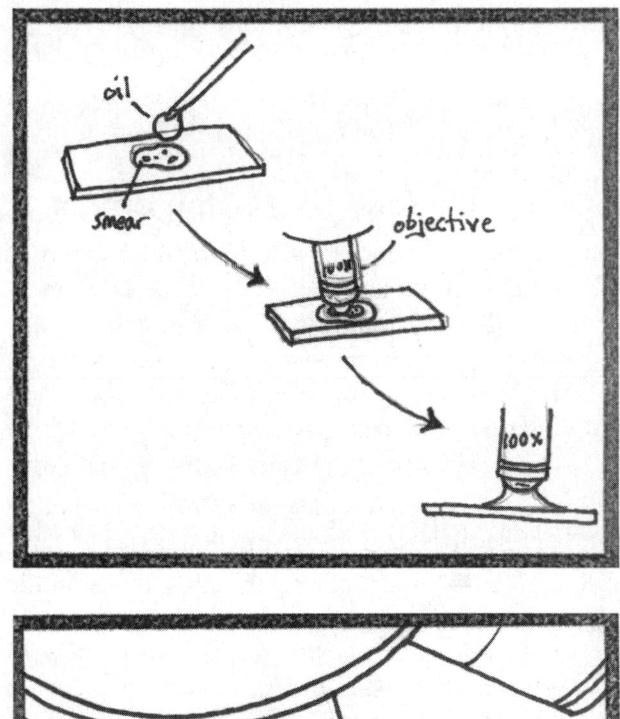

PROCEDURES

PPE and Lab Safety Review

1. Make sure that you are wearing the proper safety equipment for each experiment. Your lab coat and gloves should be worn at all times in the lab. Goggles should be worn during the centrifugation experiment, but they are not necessary for the microscopy experiments.
2. Your instructor will point out key safety equipment in the laboratory.

Centrifugation Exercise

1. You will receive a 15-mL conical tube containing water and diatomaceous earth. Some of the earth may have settled to the bottom. Invert to mix or vortex the sample until you have a nice colloidal suspension.
2. Weigh your test tube on a balance and mark the mass on the tube in pencil.

3. Insert the tube into a clinical centrifuge. Be sure to balance your sample with another tube that is +/− 0.5 g similar in mass.
4. Centrifuge the sample for 5 minutes at setting 3 on the centrifuge.
5. Carefully remove your tube and observe the sample. Can you identify the insoluble pellet and the liquid supernatant? Is there any earth left in suspension after the centrifugation?

Ocular Micrometer Measurement

1. Familiarize yourself with the specific setup of your microscope. Locate all of the parts mentioned in the background section, and practice adjusting the sample holder, adjusting the coarse and fine focus, increasing and decreasing the light intensity, and spinning the objective wheel. Do all of these before you put any sample on the stage.
2. Place a stage micrometer into the sample holder on the stage, and make sure that it is secure.
3. Using the lowest power objective, focus on the center of the micrometer. You should see a small, unlabeled ruler on the slide. Focus on the gradations of the micrometer, and look to see what the actual measurement is for each tick mark.
4. Looking through only the ocular with a number, ocular micrometer, determine the number of tick marks on the stage micrometer for the space between 0 and 1 on the ocular micrometer. Then, do the same for the space between 4 and 5 on the ocular micrometer. Convert this information into microns per tick on the ocular micrometer; then compare and average your findings and record them in the space provided in question 5 on the attached worksheet.
5. Repeat this for each of the dry objectives.
6. Lower your stage to the bottom setting. Add a drop of oil or water to the center of the micrometer, and slowly raise the stage until the oil/water objective just touches the droplet. Focus using the fine objective adjustment, then determine and record the microns per tick mark for your oil objectives, if your microscope has them.
7. When you are finished, carefully clean the micrometer slide with lens paper and return it to your instructor.

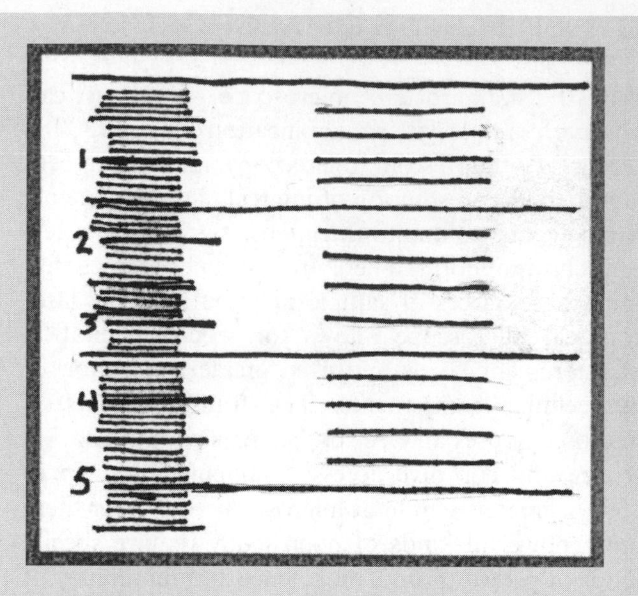

Kohler Illumination Setup

1. Place your *Zea mays* cross-section slide into the sample holder. Switch to the lowest dry objective and focus on the sample.
2. Close the iris diaphragm all the way. Change the height of the condenser lens until you can see a sharp image of the iris (in most cases either a hexagon or an octagon).
3. Center the iris using either the small screws on the illumination source or on the condenser lens.
4. Slowly open the condenser lens until the iris image is opened to fill the space just beyond the entire field of view.
5. Switch to the next dry objective and repeat the process.
6. Lower your stage to the bottom setting. Add a drop of oil or water to the slide and slowly raise the stage until the oil/water objective just touches the droplet. Focus using the fine objective adjustment, and then repeat the process for each of the oil/water objectives.
7. Every time you switch objectives you may need to slightly adjust the Kohler illumination to achieve the best, focused illumination. However, I usually only make these adjustments when I start using the microscope and right before I take any microphotographs.

Observation of Fresh Mount Slides

1. Prepare a fresh mount slide using petroleum jelly to create a small circular well on a glass slide. Add a drop of Euglena culture to the center of the well, then place a cover slip over the drop.
2. Observe the motion of the Euglena under all dry objectives.
3. Carefully remove the cover slip, and add one drop of Protoslo to the drop of culture and mix with a toothpick. Cover the sample again with a new cover slip and observe. How has the Euglena movement changed? Can you notice any internal or external structures, now that the protist's movement has been slowed? Sketch your observations for question 4 of the worksheet.
4. Now prepare a fresh mount slide using an Elodea leaf. Place the Elodea leaf down in the center of a slide. Add one drop of water, and cover with a cover slip.
5. Observe the leaf for several minutes using the dry objectives. After some time in the light, do you observe cyclosis or cytoplasmic streaming with the individual cells of the leaf? How many cell layers thick is your leaf? Sketch your observations of several Elodea cells for question 4 on the worksheet. Label the cell wall, cytoplasm, nucleus (when visible), vacuole, and chloroplasts.

WORKSHEET

1. Proper balancing of your samples is extremely important when setting up a centrifuge run. Below are diagrams of six rotors. Color in the sample holders in the pattern that you would use for the number of samples listed in the center of each rotor. Use an X for blank samples.

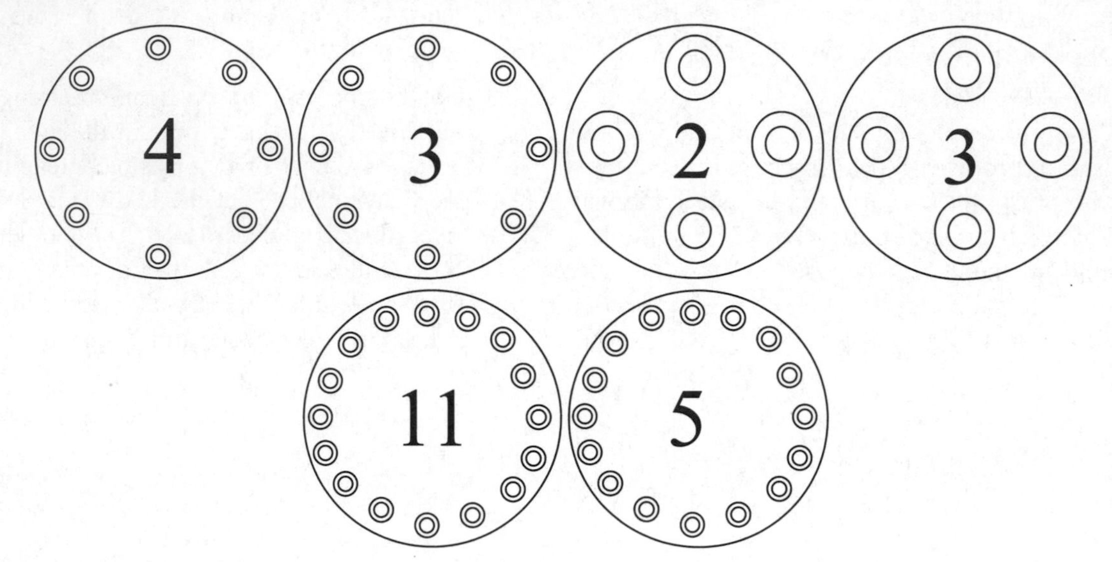

2. Did you notice any difference between the low/high dry-objectives and the oil-immersion objective? Which was clearest? Which was best for colored samples? Which was best for unstained regions of the sample? Explain why obtaining Kohler illumination for an objective is important for microscopic observations.

____

____

____

____

____

3. Draw two images of your prepared Zea stem cross-section samples, one using the low, dry-objective and one using the oil-immersion objective. Be sure to mark the magnification and add a scale bar.

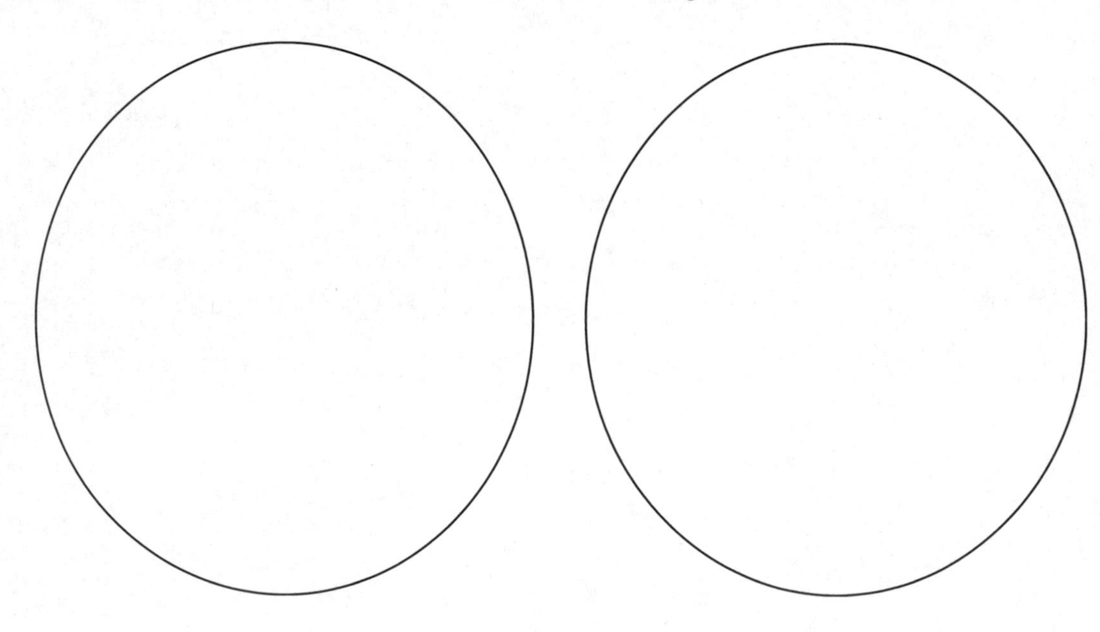

4. Below, draw representative pictures of your Elodea and Euglena wet mount samples. Be sure to mark the magnification and add a scale bar.

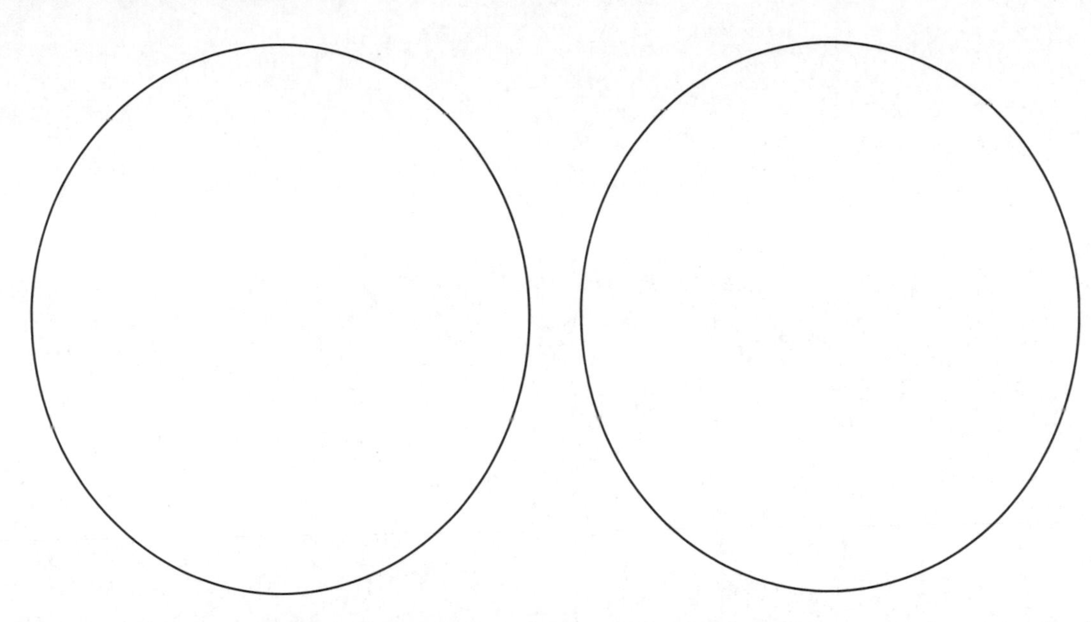

5. Below, record the ocular micrometer conversion for each objective. Be sure to keep track of which microscope you are using, since this conversion factor may differ between microscopes.

Objective	Microns per Tick

DISCUSSION

Discussion Questions

1. Personal Protective Equipment (PPE) is an essential aspect of laboratory science. Different experiments require different PPE. Discuss what types of PPE would be necessary for a Dry Lab experiment. Discuss possible PPE to wear for a histology experiment. Discuss the necessary PPE for a wet lab experiment. What PPE would you add if the wet lab experiment involved concentrated acids or bases? What if it involved volatile organic compounds?
2. Discuss the setup of a centrifugation run. What steps must be taken to ensure a safe experiment?
3. What are the limits of using compound microscopy? Discuss aspects like magnification, working distance, depth of field, and diameter of field, and how these characteristics affect sample preparation and what you can/cannot visualize using the compound microscope.
4. What is the purpose of using Kohler illumination? What are the benefits of adjusting your microscope, using this technique?
5. Today, you observed a number of wet-mounted samples. When would you use a wet mount or a "vital preparation" in an experiment? Today, you also observed a number of prepared samples, or "fixed" slides. When would you prepare fixed slides of a sample? What types of experiments would call for this histological technique?

Enzyme Kinetics

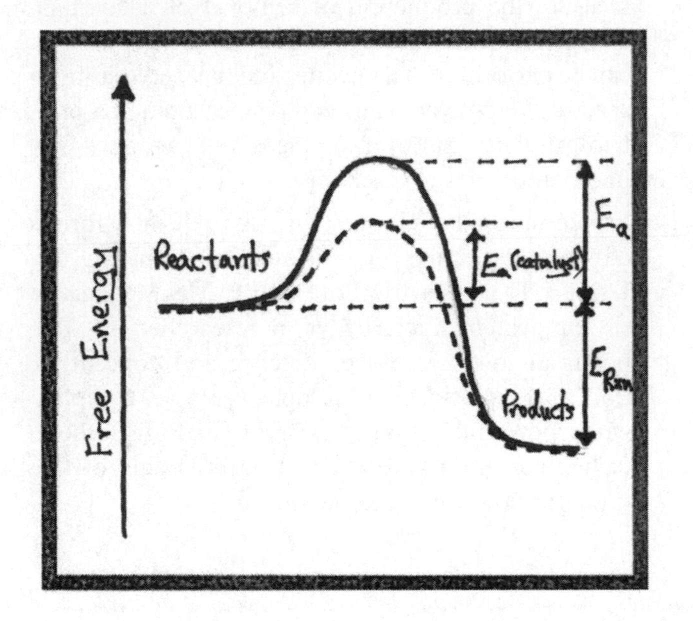

Objectives

- Isolate crude protein from bacterial cultures.
- Perform a colorimetric β-galactosidase assay on your isolates.
- Use this assay to study the regulation and activity of β-galactosidase.

DESCRIPTION

Once the structure of DNA had been proposed and accepted, the question then arose of how the genetic information contained in DNA was expressed, and how that expression was regulated. Gene expression refers to the active transcription of a gene. Clusters of genes in bacteria, called operons, can be activated all at once in quick response to a change in environmental conditions. We will be studying the *lac* operon in *E. coli*. Typically, this operon is activated in the presence of lactose in bacterial growth medium; however, it is also the target of a number of assays used throughout molecular biology. Today, you will use an enzyme activity assay to study the regulation of the production and the activity of β-galactosidase.

CONCEPTS & VOCABULARY

- Active site
- Competitive inhibitor
- ΔG
- k_{cat}
- K_{eq}
- K_M
- *lac* operon
- Non-competitive inhibitor
- V_{max}

CURRENT APPLICATIONS

- Enzyme activity assays have been a common experimental technique for over a century. The first enzyme identified was catalase, which is responsible for the breakdown of hydrogen peroxide. The activity of this enzyme was first observed as it decomposed H_2O_2 into water and O_2. The evolution of oxygen gas was detected and used to analyze the reaction rate of the enzyme. Current enzymatic assays typically measure the production or removal of a colored/fluorescent/luminescent product/substrate. Assays can be performed in a cuvette, and measured using a spectrophotometer. They can be measured in a protein gel. They can even be measured by naked-eye observation of a color change.
- Proteomic analysis is one of many of the new forms of robust molecular analysis available to biologists. Using 2-D gel electrophoresis, LC-MS-MS analysis, or protein microarrays, a researcher has the potential to observe the presence and concentration of all proteins in a biological sample. Coupled with powerful computer-based analysis, these techniques are a gateway to understanding of the protein nature of the proteome.

REFERENCES

Karp, G. (2010). Enzymes as biological catalysts. *Cell and Molecular Biology: Concepts and Experiments.* (92–104) Hoboken, NJ: John Wiley & Sons, Inc.

Sambrook, J, Fritsch, EF, and Maniatis, T. (1989). Beta-galactosidase assays. *Molecular Cloning: A Laboratory Manual* 2nd ed., (B. 14, p. 186). Cold Spring Harbor: Cold Spring Harbor Laboratory Press.

BACKGROUND

ENZYMES

Enzymes are biological catalysts. A catalyst is a compound that can accelerate the rate of reaction but that is not altered by the reaction itself. Catalysts have no effect on the thermodynamics of a reaction, as they do not supply energy to the reaction itself. That is to say, they do not affect the ΔG but they do affect the K_{eq}.

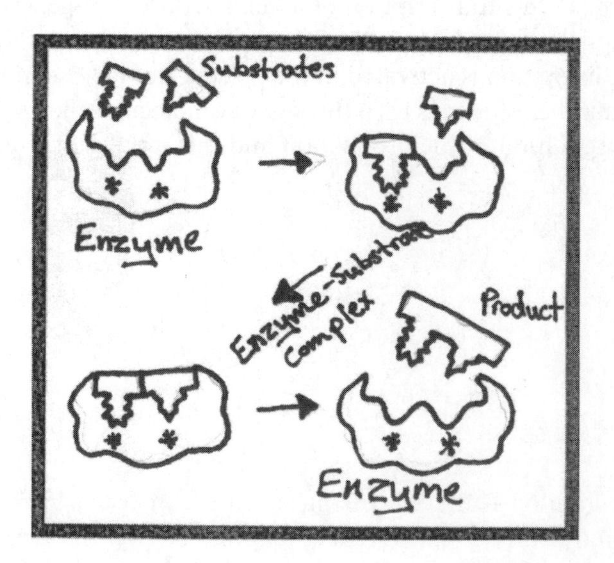

STRUCTURE AND MECHANISM OF ENZYME ACTIVITY

Enzymes are protein catalysts. Typically, they consist of one or more proteins bound to a cofactor. These cofactors may be metal ions or organic coenzymes. Whereas organic catalysts may increase the rate of the

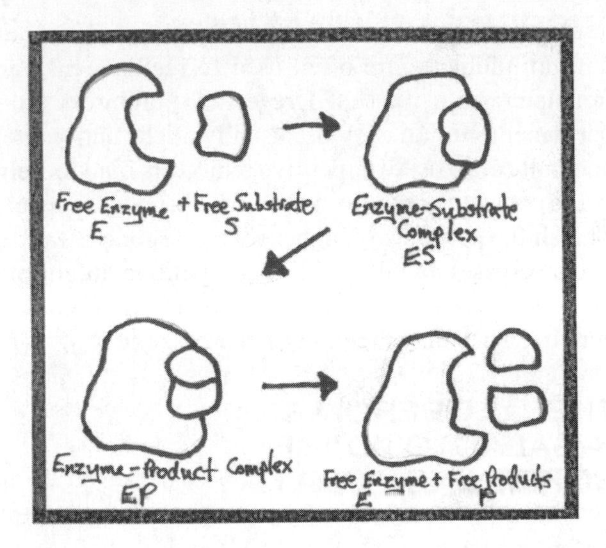

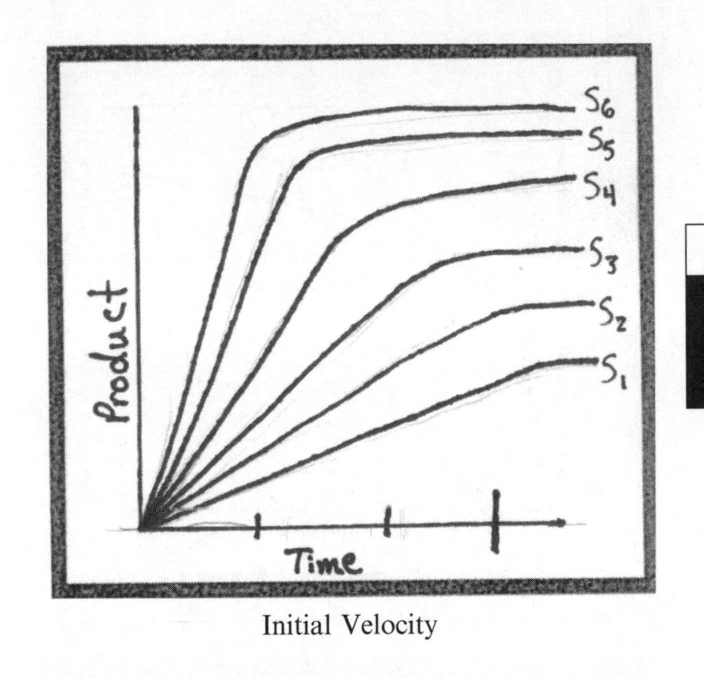

Initial Velocity

reaction 1000-fold, enzymes typically increase the rate 10^8- to 10^{13}-fold. This is achieved by reducing the energy of activation. The energy of activation is the energy required to push the reactants into a transition state, which will then form the product compound. The enzyme achieves this by forming an enzyme-substrate complex, in which the reactants are bound into the active site of the enzyme. Once this complex is formed, the enzyme reduces the energy of activation by substrate organization, changing the substrates' electrostatic properties, and/or exerting physical stress on bonds within the substrate. The most amazing thing about enzymes is that they can achieve this at body temperature and at physiological pH.

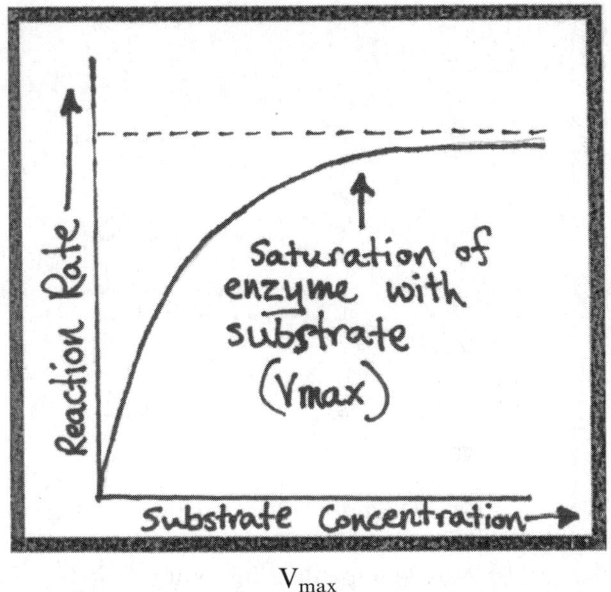

V_{max}

ENZYME KINETICS

Enzyme kinetics deals with the rate of enzyme catalysis under various conditions. The rate of reaction or velocity is calculated for varying concentrations of various substrates. The V_{max}, or maximum velocity, is the velocity at saturation. Saturation refers to the concentration of substrate where every enzyme molecule is occupied by substrate at a given time. The k_{cat}, or turnover number, is the maximum number of substrate molecules that can be converted to product by one enzyme every second. Values of one to one-thousand are common for most enzymes. The Michaelis constant or K_M represents the affinity of enzyme for substrate. The K_M can be calculated using the Michaelis-Menten. K_M values for most enzymes of 0.1 to 0.0001 are common. Determination of this value is helpful in the study and utilization of an enzyme.

Energy of Activation

INHIBITION

Inhibition is the process in which a compound may bind to an enzyme to reduce its productivity. Enzyme

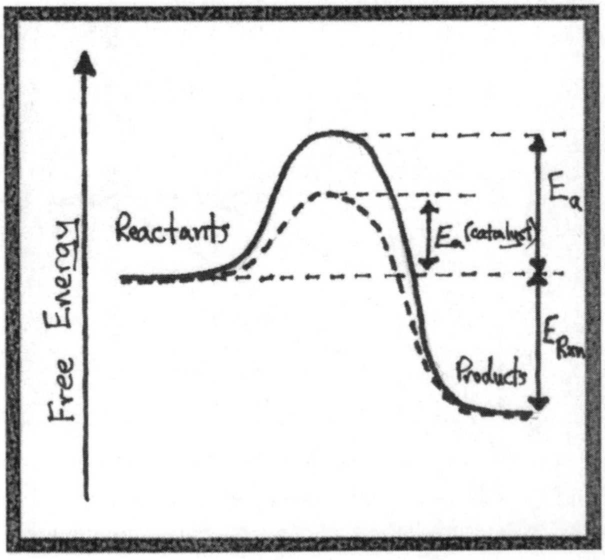

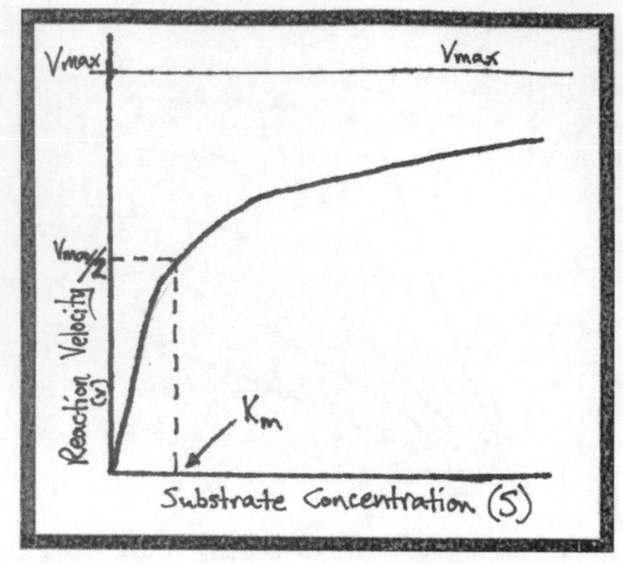

Enzyme Kinetics

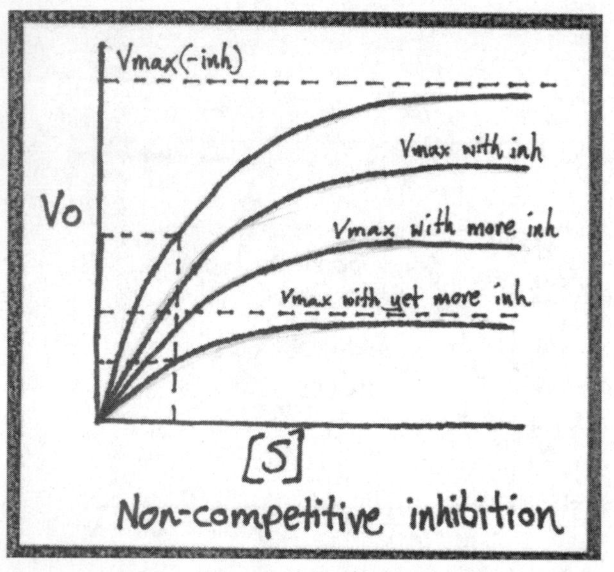

Non-Competitive Inhibition

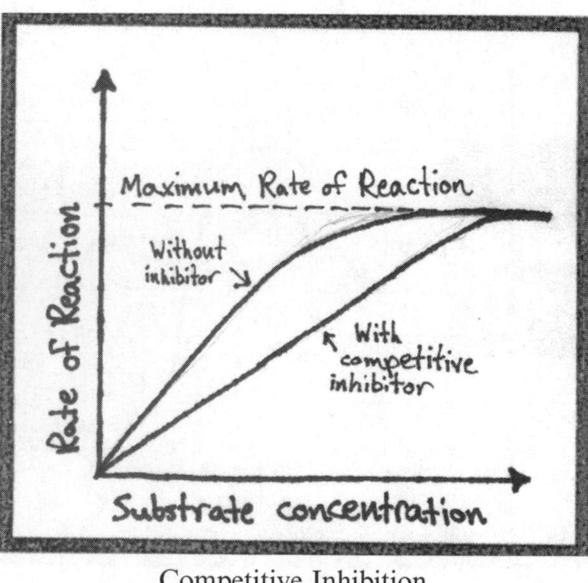

Competitive Inhibition

inhibitors often take the form of drugs or toxins, but natural inhibitors are often used to regulate enzyme function within the cell. Irreversible inhibitors bind permanently to an enzyme, usually at the active site, and halt catalysis. Competitive inhibitors bind loosely to the active site and compete with substrate for space. This shifts the V_{max} to a higher substrate concentration and decreases the K_M. Non-competitive inhibitors bind to another site on the enzyme and alter enzyme activity. They may increase or decrease the K_M.

THE *lac* OPERON & β-GALACTOSIDASE SCREENING ASSAY

The *lac* operon is a family of bacterial genes that are involved in the breakdown of the carbohydrate lactose. In the presence of glucose, these genes are not expressed. However, in the absence of glucose and the presence of lactose, this entire family of genes is induced. The expression of these genes is controlled by an inducible operon controlled by the binding of a repressor protein. In the absence of lactose, the repressor protein binds to the operator sequence and prevents transcription by blocking RNA polymerase from binding to the promoter sequence. In the presence of lactose, the repressor protein is inactivated by the sugar. So long as there is available lactose to bind the repressor protein, it is inhibited from binding to the operator sequence. If the repressor protein cannot bind to the operator sequence, RNA polymerase binds to the promoter sequence of the operon and begins expression of the structural genes that follow. These genes are involved in the breakdown of lactose, but are only expressed when lactose is the most available source of carbohydrate.

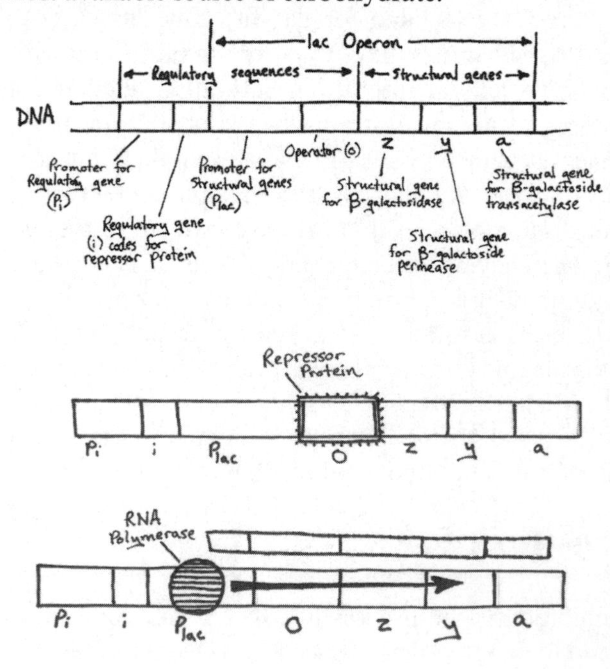

Today you will be performing the β-galactosidase enzyme assay on bacterial cell extracts. The bacterial cultures will be grown in the presence or absence of IPTG (a synthetic lactose mimic that permanently binds the repressor protein) and glucose. The assay uses the colorimetric product from the breakdown of ONPG, a synthetic lactose mimic. ONPG forms a yellow product when catalyzed by β-galactosidase. You will determine the kinetics of this enzyme in order to expand your understanding of the topic.

PROCEDURES

Measurement of Bacterial Density and Lysis of Bacterial Cells

1. The four cultures have been grown overnight and placed into flasks labeled BLANK, A, B, and C.
2. Turn on your spectrophotometer, and set it for absorbance at 595 nm.
3. Measure the A_{595} for each culture. Record the data on the worksheet for question 1.
4. Get four 5-mL test tubes and label them: BLANK, A, B, and C.
5. Place 2.0-mL of the appropriate bacterial suspension into each tube.
6. In the hood, add 1.0-mL of chloroform and 0.6-mL of 0.10% sodium sarcosyl to each tube to lyse the bacteria and release the enzyme. Vortex for 10-seconds on the highest setting.

Measurement of β-Galactosidase Activity

1. Switch the spectrophotometer setting to absorbance at 420 nm.
2. Label 9 cuvettes Blanks 1, 2, and 3, and Samples 1 through 6. Remember to keep all reaction mixtures at room temperature, but keep your bacterial lysate on ice.
3. Make Sample 6 first, and incubate it in a boiling water bath for 5 minutes; then put it on ice.
4. Add the assay medium to each of the remaining cuvettes.
5. Create the Blanks by adding the appropriate amount of the blank culture lysate to the assay medium. Cover with Parafilm and invert 4-times to mix.
6. Move to the spectrophotometer.
7. Adjust for Blank 1.
8. Add the appropriate amount of bacterial lysate to Sample 1, cover with Parafilm, and invert 4-times to mix.
9. Measure the sample for the 0-minute data-point immediately.
10. Adjust for Blank 2.
11. Add the appropriate amount of bacterial lysate to Sample 2, cover with Parafilm, and invert 4-times to mix.
12. Measure the sample for the 0-minute data-point immediately.
13. Adjust for Blank 3.
14. Add the appropriate amount of bacterial lysate to Sample 3, cover with Parafilm, and invert 4-times to mix.
15. Measure the sample for the 0-minute data-point immediately.
16. Repeat steps 14 and 15 for Samples 4–6.
17. Set the cuvettes aside at room temperature.
18. Measure all of the samples every 2-minutes for another 10-minutes. Remember to adjust for Blank 1, then measure Sample 1; adjust for Blank 2, then measure Sample 2; and adjust for Blank 3, then measure Samples 3–6.
19. Collect your data on the datasheet for question 1.

Tube	IPTG	Glucose	Assay Medium	Bacterial lysate
Blank 1	+		0.3-mL	0.3-mL*
Sample 1	+		0.3-mL	0.3-mL**
Blank 2	+		0.9-mL	0.9-mL*
Sample 2	+		0.9-mL	0.9-mL**
Blank 3	+		0.6-mL	0.6-mL*
Sample 3	+		0.6-mL	0.6-mL**
Sample 4		+	0.6-mL	0.6-mL***
Sample 5			0.6-mL	0.6-mL†
Sample 6	+		0.6-mL	0.6-mL**

* Use Blank Culture lysate for these samples.
** Use Culture A.
*** Use Culture B.
† Use Culture C

WORKSHEET

1. Write down the absorbance for each tube in the space below

Tube	0-minutes	2-minutes	4-minutes	6-minutes	8-minutes	10-minutes
Sample 1						
Sample 2						
Sample 3						
Sample 4						
Sample 5						
Sample 6						

Now calculate the change in absorbance between each time point between 0 and 10-minutes.

Tube	ΔAbs 0–2 m	ΔAbs 2–4 m	ΔAbs 4–6 m	ΔAbs 6–8 m	ΔAbs 8–10 m
Sample 1					
Sample 2					
Sample 3					
Sample 4					
Sample 5					
Sample 6					

On the graph paper provided, plot the ΔAbs versus time. Use 0 as a ΔAbs for 0-minutes. Draw a best-fit curve with a ruler and label each line with the Sample number.

2. Estimate the initial velocity by using the change in Absorbance for each sample between 0 and 2-minutes. Which volume of Sample gives the highest initial velocity? Explain why?

____

____

____

____

____

____

3. For Samples 1–3, how does the change in the amount of suspension change the initial velocity of the reaction? Is this a linear change as the amount doubles, the rate doubles? Explain your answer.

____

____

____

____

____

____

4. Sample 4 contained glucose, a non-competitive inhibitor of β-galactosidase. How does the initial velocity of Sample 4 differ from Sample 3? Why is this the case?

____

____

____

____

____

5. Sample 5 did not contain IPTG. IPTG permanently binds the *lac* operon repressor protein. How does the initial velocity of Sample 5 differ from Sample 3? Why is this the case?

____

____

____

____

____

6. Sample 6 contains a heat-treated bacterial lysate. How does the initial velocity of Sample 6 differ from Sample 3? Why is this the case?

____

____

____

____

____

DISCUSSION

1. Discuss why β-galactosidase was chosen for this assay, instead of another enzyme from the *lac* operon.
2. Discuss why growth on IPTG was included in some of the samples tested by the assay. How would the removal of this compound from the growth media change the data collected?
3. Glucose is a non-competitive inhibitor for the β-galactosidase. Discuss how this sugar inhibits β-galactosidase expression. How would the addition of this compound change the data collected?
4. What would heat-treatment in boiling water do to the proteins in the bacterial isolate? How would this treatment change the data collected?
5. The β-galactosidase in your samples will consume the available ONPG as the reaction progresses. The activity of β-galactosidase will slowly decrease in a bacterial isolate at room temperature. How will these two factors change the data collected over a 30-minute time course?

OVERVIEW

3 Membrane Permeability

Objectives

- To observe plasmolysis and deplasmolysis of Elodea cells.
- To test the rate of penetration of various alcohols into Elodea cells.
- Determine what molecular characteristics increase or decrease penetration.

DESCRIPTION

All cells are encompassed by a phospholipid bilayer embedded with proteins and other lipids/sterols. Membranes are permeable to small, non-polar, organic molecules. Larger, more-polar, or charged molecules and ions move more slowly through the membrane, if at all. These molecules are often shuttled through the membrane using a variety of substrate-specific transport proteins. These proteins may offer a channel for specific or non-specific, facilitated diffusion, or they may utilize ATP to perform active transport of these molecules into the cytosol. Plant cells contain a protoplast surround by a cell wall. The protoplast can separate from the cell wall due to water loss. This process is called plasmolysis. Through observation of the rate of plasmolysis/deplasmolysis, one may determine the nature of the plasma membrane's permeability, the penetration rate of various organic molecules.

CONCEPTS & VOCABULARY

- Active transport
- Diffusion
- Ether: water partition coefficient
- Facilitated diffusion
- Fluid mosaic molecule
- Osmosis
- Osmolarity
- Plasmolysis/deplasmolysis
- Passive transport
- Protoplast
- Tonicity

CURRENT APPLICATIONS

- Plant cells are designed to perform at peak efficiency while they are in a swollen state of turgor. Plant cells found at conditions considered isotonic for an animal will have reduced enzyme activity and nutrient uptake.
- The cell walls of plant cells can be removed using enzymatic digestion. This can be used to create separated, single cells called protoplasts. Protoplasts can be used to study plant cells without using whole plants.
- Water potential in cells is regulated by solute concentration and pressure on the cell membrane. In plant cells, solutes are stored in the vacuole and removed or released into the cytosol to balance water potential. The standard permeability of these membranes, coupled with transport proteins found in the membrane, is essential to maintain this regulation.

REFERENCE

Karp, G. (2010). Movement of substances across cell membranes; endocytosis. *Cell and Molecular Biology: Concepts and Experiments*. (143–158, 297–299). Hoboken, NJ: John Wiley & Sons, Inc.

BACKGROUND

FUNCTIONS OF THE PLASMA MEMBRANE

Above all other organelles, the plasma membrane is an essential characteristic of a cell. Cells may lack mitochondria, a definitive endomembrane system, and even a nucleus, and still be considered a cells, so long as they are enclosed in a plasma membrane. The most basic function of the plasma membrane is to separate the interior of the cell from external environment. Nutrients, ions, and dissolved gasses may exist at varying concentrations within the external environment. But within the cytoplasm of a cell protected by an intact cell membrane, these molecules may be concentrated or diluted in order to meet the requirements to maintain homeostasis.

The plasma membrane provides a semi-permeable barrier. The membrane allows the passage of hydrophobic organic compounds, and water to a limited extent. These chemicals can freely pass through the membrane via simple diffusion or osmosis. The membrane is also maintained at a net-negative charge in order to attract positive ions. The amphipathic nature of the plasma membrane allows for the integration of protein and lipid/sterol components. The specific structure of a region of the plasma membrane has a profound effect on its function.

STRUCTURE AND COMPONENTS OF THE PLASMA MEMBRANE

Most plasma membranes are made of a phospholipid bilayer, with a carbohydrate head pointing out toward the aqueous side and the hydrophobic tails, pointing inward toward each other. This creates a hydrophobic region in between the two layers of the membrane. The membrane is not uniform in its composition, as other lipids may be mixed throughout. These lipids may be involved in regulatory processes, such as cell-to-cell recognition or signal transduction. These lipids may also be modified with carbohydrates. Lipids with different physical properties may also serve some other function. For example, cholesterol and stearic acid both remain solid at body temperature. These lipids may stiffen the membrane to offer mechanical support. Other lipids with a lower melting temperature, like linolenic acid, can help to maintain membrane fluidity at low temperatures and prevent damage due to freezing.

The protein components of a plasma membrane are also involved in cell-to-cell recognition and signal transduction. They, too, may be modified with carbohydrates, but they may also be modified with various lipids. The protein components of the plasma membrane are primarily involved in transport of materials across the membrane. Ions and compounds would freely diffuse across the plasma membrane due to a concentration difference, but cannot pass the membrane or enter or exit the cell via facilitative diffusion. Facilitated diffusion requires a pore or a gate protein that traverses the membrane. These protein complexes can simply be holes that allow for free passage of materials, but they are typically specific to one type of ion or compound. To move a molecule across the plasma membrane against its concentration gradient requires active transport. Active transport proteins pump molecules from one side of the membrane to the other fueled by ATP. These proteins are involved in the concentration of essential ions and nutrients inside the cell, as well as in maintaining the negative electric potential associated with the plasma membrane. The activity and concentration of many of these transport proteins is tightly regulated and can change in response to various stimuli or stresses. One last group of proteins associated with the plasma membrane are the components of the cytoskeleton. These proteins provide structural support, changes in membrane morphology, and large-scale movement of materials into and out of the cell.

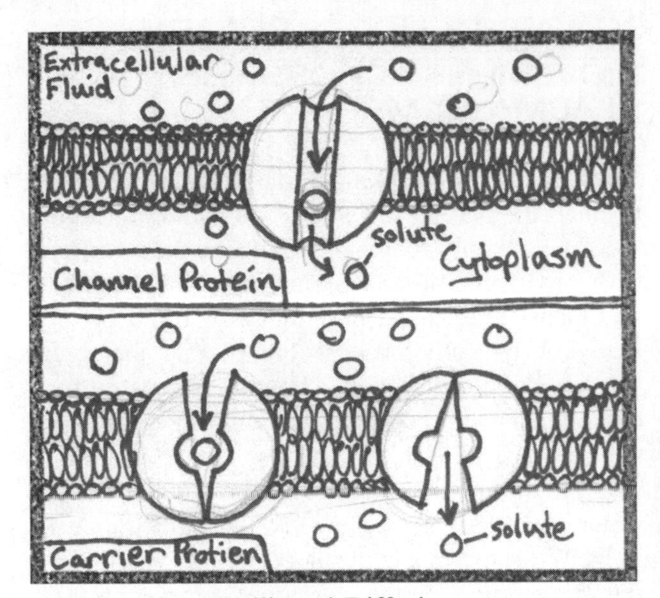

Facilitated Diffusion

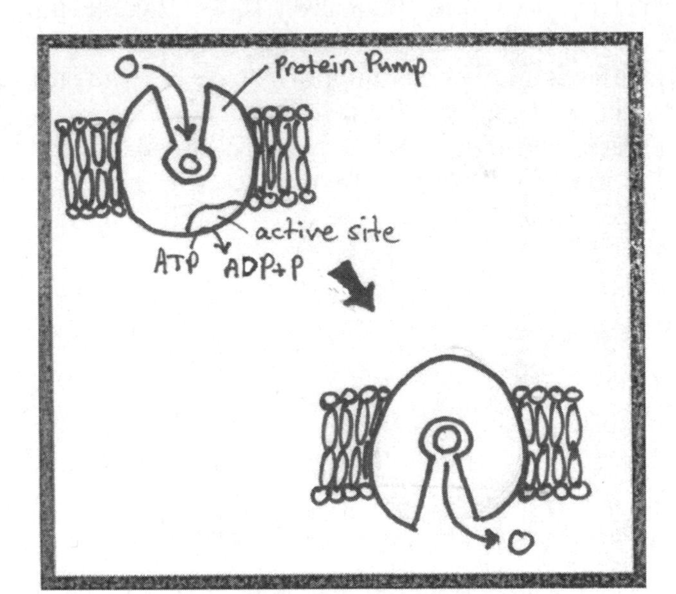

Active Transport

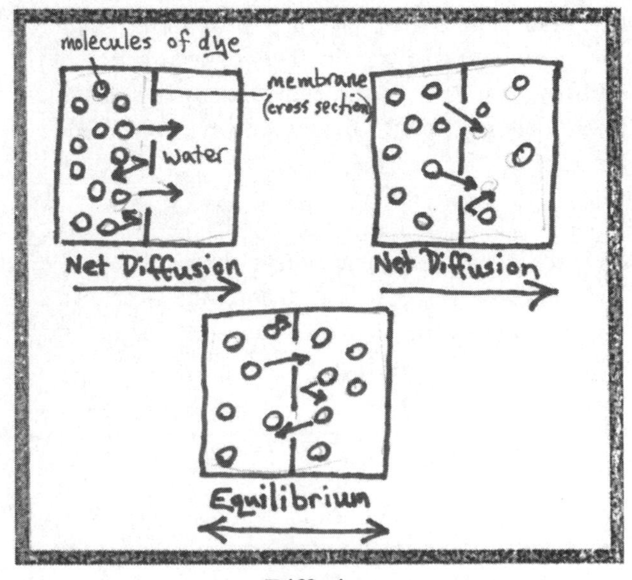

Diffusion

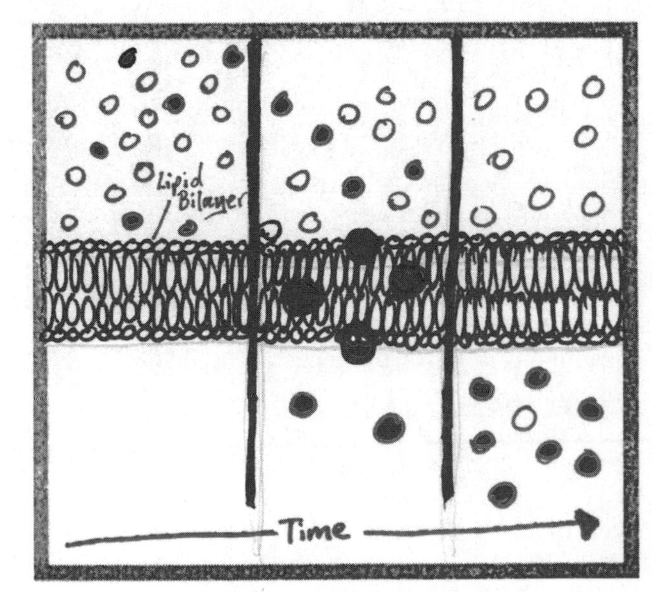

Semi-Permeable Membrane

PERMEABILITY AND SOLUTE MOVEMENT ACROSS THE PLASMA MEMBRANE

As we have discussed, water molecules are able to pass through the plasma membrane via osmosis. Electrolytic ions are only able to diffuse across the membrane by facilitative or an active transport. Nonelectrolytes and organic molecules may require a transport protein, but typically traverse the plasma membrane through passive diffusion. The rate of penetration is determined by the compounds' solubility in lipids and the presence of charged groups. Lipid solubility is calculated using the ratio of its solubility in ether and water—for example, the ether/water partition coefficient. While organic molecules typically pass through the membrane, certain organic compounds may move more slowly. Take alcohols, for example; the more organic the alcohol (linked to carbon number) the faster the rate of diffusion. This increases with each additional carbon. However, additional hydroxyl groups would decrease the rate of diffusion. Smaller molecules will also diffuse faster than larger molecules. Today, in this experiment, you will use the rate of plasmolysis to deplasmolysis in Elodea leaves to estimate the rate of penetration of a number of different alcohols. You will use your observations as an example to discuss how different molecules enter the cell.

3

OVERVIEW

Ether:water partition coefficients (solubility in diethyl ether)/(solubility in water) for a series of alcohols. The alcohols are grouped according to the number of carbon atoms and are arranged within each group by increasing number of hydroxyl groups.

Alcohol	*Condensed structural formula*	*Partition coefficient*
Methanol	CH_3OH	0.14
Ethanol	CH_3CH_2OH	0.26
Ethylene glycol	$HOCH_2CH_2OH$	0.0053
1-Propanol	$CH_3CH_2CH_2OH$	1.9
Propylene glycol	$CH_3CH(OH)CH_2OH$	0.018
Glycerol	$HOCH_2CH(OH)CH_2OH$	0.00066

PROCEDURE

Plasmolysis and Deplasmolysis

1. Cut an Elodea leaf and drop into two drops of 0.6M sucrose solution and cover with a coverslip.
2. Observe and sketch the plasmolyzed cell for question 1.
3. Carefully remove the coverslip. Drain the sucrose solution by using a piece of paper towel to wick the liquid away.
4. Add 2 drops of dd-water and cover with a new coverslip.
5. Observe the cell. Note the amount of time it takes the majority of cells to de-plasmolyze.
6. Observe and sketch the plasmolyzed cell for question 1.

Measurements of the Relative Rates of Penetration of a Series of Alcohols in Elodea

The following alcohols are to be tested: methanol, ethanol, ethylene glycol, 1-propanol, propylene glycol, and glycerol. The concentration of each is 0.4M in an isotonic sucrose solution. Each alcohol should be tested on a fresh Elodea leaf.

1. Place a cut Elodea leaf onto a clean microscope slide. Blot the leaf dry. Record the time. Add two drops of the methanol-sucrose solution and cover with a coverslip.

2. Examine the cells of the leaf under high-dry magnification. Observe for 5-minutes before recording. Determine the extent of plasmolysis (none, 25%, 50%, 75%, 100%).
3. Observe for another 5-minutes for any de-plasmolysis. Record the time.
4. Repeat these three steps for each alcohol-sucrose solution and determine the rate of penetration for each alcohol. Organize the alcohols from fastest to slowest in the table in question 2.

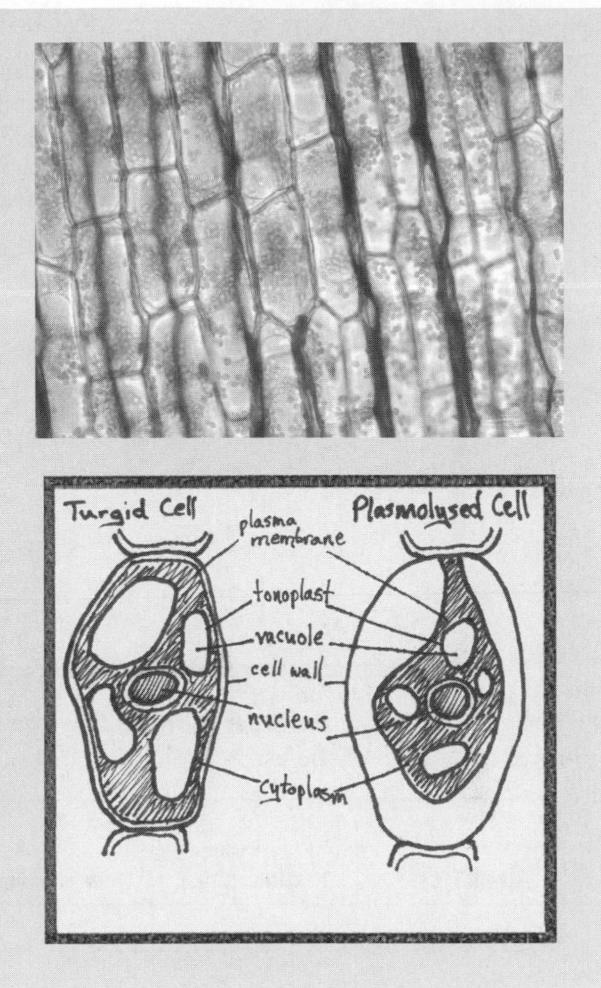

WORKSHEET

1. Draw examples of the plasmolyzed and deplasmolyzed Elodea cells. Describe any visible structures, and draw an example sketch in the space provided below. Remember to label all microscope images with the following: organism, magnification, and treatment, and add a scale bar at the bottom right of each sketch.

3 OVERVIEW

2. List the alcohols in the order in which they penetrate the plasma membrane of Elodea leaf cells, the most rapid at the top of the list.

Alcohol	Ether: water Partition Coeff	# carbons	# -OHs	Molecular Size (1–6, 1=smallest)	Rate of penetration (1=fastest)

3. Define passive and active transport in cells. Give examples of each in reference to the plasma membrane. Explain the difference between diffusion and osmosis. What is diffusion dependent upon? What is osmosis dependent upon?

4. Describe the relationship between a molecule's ether, water partition coefficient, and its rate of penetration.

5. Describe the relationship between a molecule's molecular size and its rate of penetration. Describe the relationship between a molecule's number of carbons and its rate of penetration. Describe the relationship between a molecule's number of hydroxyl groups and its rate of penetration.

3 OVERVIEW

DISCUSSION

1. Discuss the structure of the plasma membrane. How does the lipid composition affect permeability? What are the functions of the protein components? Of the carbohydrate components?

2. Discuss the hyper-, hypo- and isotonic solutions. What happens what you place an animal cell into each of these solutions? What happens to a plant cell? Explain your answers.

3. Discuss how polar molecules pass through the plasma membrane. What increases or decreases this movement?

4. Discuss how non-polar molecules pass through the plasma membrane. What increases or decreases this movement?

5. Define facilitated diffusion. Define active transport. What is required for these processes to exist at the plasma membrane of cells?

OVERVIEW 4

Protein Transport to the Plasma Membrane

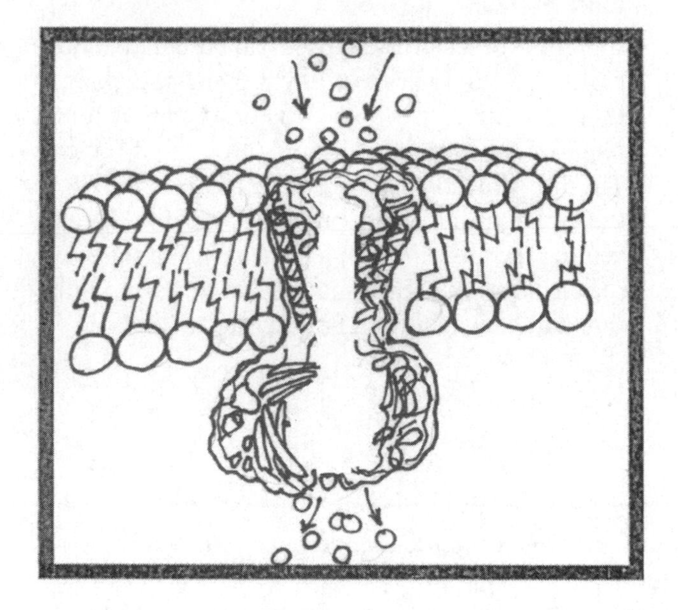

Objectives

- Isolate plasma membrane and associated proteins from red blood cells (i.e. erythrocyte "ghosts").
- Determine the activity of the ouabain-sensitive Na+/K+-ATPase found in the erythrocyte plasma membrane.

DESCRIPTION

The plasma membrane is a contiguous phospholipid bilayer punctuated by protein, sterol, and other modified lipid components. While the semi-permeable bilayer allows for movement of water, certain ions, and most organic molecules, it is the protein components that allow for the trafficking of most molecules into or out of the cell. There are three basic types of plasma membrane proteins: integral, peripheral, or lipid-anchored. These proteins are synthesized at the rough endoplasmic reticulum (RER) via the ribosomes, then modified and packaged for transport into vesicles at the trans-Golgi network. Secretory vesicles are transported to the plasma membrane. In order to identify and study these proteins, specific extraction techniques are used. Today, you will be isolating plasma membranes from erythrocytes and studying the activity of the plasma membrane protein: Na+/K+-ATPase.

CONCEPTS & VOCABULARY

- Buffy coat
- Erythrocyte ghosts
- Integral membrane proteins
- Lipid-anchored membrane proteins
- $Na+/K+$-ATPase
- Ouabain
- Peripheral membrane proteins
- Secretory pathway

CURRENT APPLICATIONS

- Isolation of plasma membranes from other cell types is more difficult than the method you will use today. In order to remove the endomembrane system and other organelles, a more complicated centrifugation involving a density gradient, the initial modification of the plasma membrane, or column chromatography. These methods will provide pure plasma membranes for use in assays or other proteomic analyses.
- Inhibitors of secretory transport are useful to the cell biologist who wishes to study protein trafficking in the cell. An active and dynamic cytoskeleton is required for protein transport throughout the cell. The drug brefeldin A prevents traffic from the ER to the Golgi, also altering protein trafficking. By preventing the movement of a protein within the cell at certain locations, information about its modification and packaging can be determined.

REFERENCES

Karp, G. (2010). The structure and function of the plasma membrane; types of vesicle transport and their functions. *Cell and Molecular Biology: Concepts and Experiments.* (117–165; 288–297). Hoboken, NJ: John Wiley & Sons, Inc.

Harris, JR, Graham, J, Rickwood, D. (2006). Purification of human erythrocyte "ghosts"; Assay for ouabain-sensitive $Na+/K+$-ATPase. *Cell Biology Protocols.* (137–138, 144). Hoboken, NJ: John Wiley & Sons, Inc.

BACKGROUND

PROTEIN COMPONENTS OF THE PLASMA MEMBRANE

The plasma membrane is a contiguous phospholipid bilayer punctuated by protein, sterol and other modified lipid components. While the semi-permeable bilayer allows for movement of water, certain ions, and most organic molecules, it is the protein components that allow for the trafficking of most molecules into or out of the cell. There are three basic types of plasma membrane proteins. Cell membrane proteins may be integral, fully spanning the membrane; they may be peripheral, associated by non-covalent bonding with a hydrophilic surface on either side of the membrane; or they may be lipid-anchored, still located on one of the hydrophilic surfaces of the membrane, but bound covalently to the a lipid molecule.

In order to identify and study these proteins, specific extraction techniques are used. Integral membrane proteins have one or more hydrophobic regions, called transmembrane domains, which help to insert and anchor the protein into the membrane. Integral membrane proteins can be separated from membranes using strong detergent treatment at high temperature and centrifugation. Since peripheral membrane proteins are only associated with the exterior of the membrane, they can be separated from membranes using high salt treatment (to disrupt electrostatic bonding) followed by centrifugation. Lipid-anchored membrane proteins can be separated from membranes by detergent treatment and centrifugation, but this process will extract the integral membrane proteins, as well. This result can be avoided by performing a shorter extraction, using detergent treatment at low temperature and centrifugation.

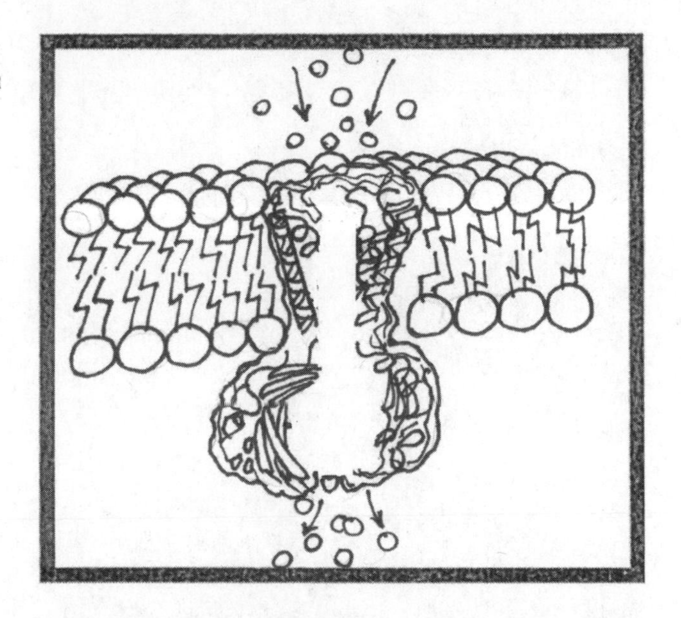

THE SECRETORY PATHWAY

Proteins are synthesized at the rough endoplasmic reticulum (RER) via the ribosomes. Proteins bound for secretion outside of the plasma membrane are synthesized into the lumen of the ER. Membrane-bound proteins are inserted into the ER membrane. Newly-synthesized proteins are modified in the ER and Golgi apparatus, if needed, and then packaged for transport into vesicles at the trans-Golgi network. Secretory vesicles are transported to the plasma membrane. The vesicles fuse with the plasma membrane and either merge vesicle membranes and integral proteins with the surrounding cell membrane, or release their contents into the extracellular space. Peripheral or lipid-anchored proteins on the exterior of the plasma membrane are secreted, and then bind to the exterior. Peripheral or lipid-anchored proteins on the interior of the plasma membrane are synthesized into the cytoplasm and then transported to the cytoplasmic face of the membrane.

There are two types of secretion: constitutive and regulated. Constitutive secretion is ubiquitous. So long as the secreted protein is being expressed and processed, it will be constantly secreted. Most secreted proteins are constitutively secreted. Regulated secretion is a controlled process. These proteins are stored in specialized vesicles bound to the plasma membrane and only released after a secondary signal is received. Proteins that require tighter control use regulated secretion. Examples include hormones, neurotransmitters, and digestive enzymes.

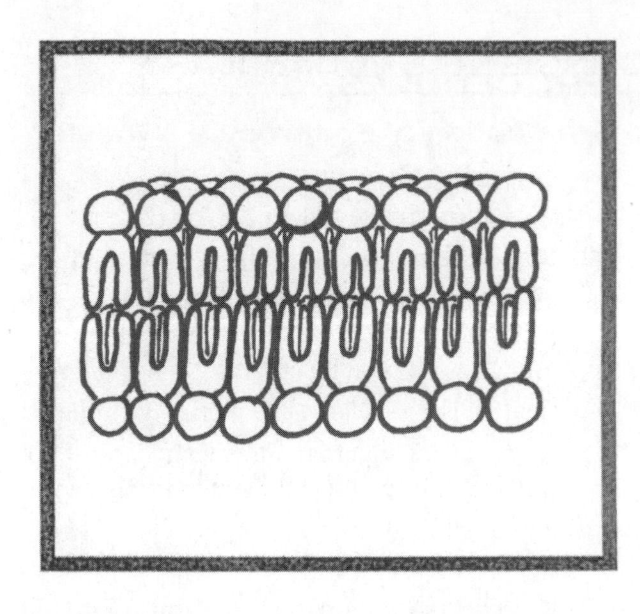

INHIBITION OF THE SECRETORY PATHWAY

There are a number of drugs and toxins that inhibit secretion in the cell. There are drugs that can stop protein synthesis. There are drugs that effect the reorganization, and motors of microtubules or microfilaments can prevent movement of vesicles away from the trans-Golgi network, or fusion of the vesicles with the plasma membrane. Brefeldin A (BFA) inhibits protein transport from the ER to the Golgi. These drugs are used to dissect the transport pathways of membrane-bound proteins.

Today, you will be isolating plasma membranes from erythrocytes and studying the activity of the plasma membrane protein: Na+/K+-ATPase. Erythrocytes are ideal cells for the isolation of plasma membranes because these cells contain no nuclei and only a simple complement of other organelles. We will also use an inhibitor of Na+/K+-ATPase to determine the sensitivity of our assay.

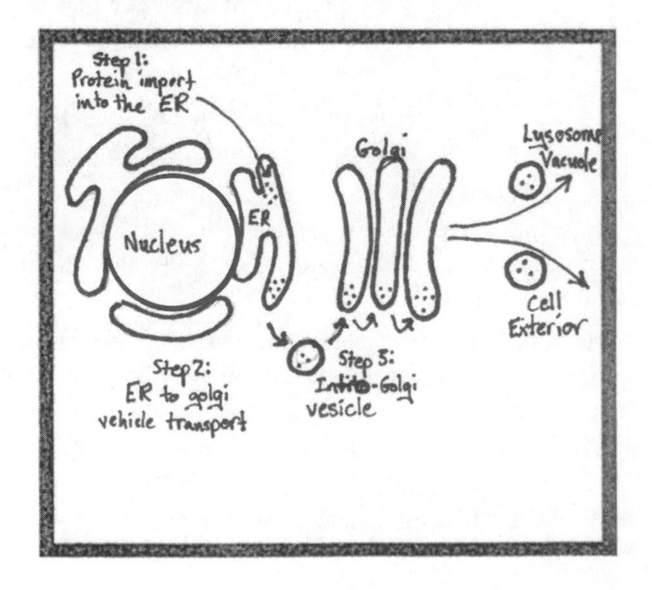

PROCEDURES

Purification of Erythrocyte "Ghosts"

1. Centrifuge your whole blood sample in a clinical centrifuge at 800 ×g for 15 minutes.
2. Remove the plasma and buffy coat, and dispose of them in the biohazard waste.
3. Add 5-mL of Buffer A and resuspend the cells using a wide-bore pipette.
4. Centrifuge the cells at 800 ×g for 15 minutes.
5. Remove the supernatant, and add 5-mL of Buffer A, and resuspend the cells using a wide-bore pipette.
6. Centrifuge the cells at 800 ×g for 15 minutes.
7. Remove the supernatant, and add 5-mL of Buffer A and resuspend the cells using a wide-bore pipette.
8. Centrifuge the cells at 800 ×g for 15 minutes.
9. Remove the supernatant, and add 30-mL of Buffer B and resuspend the cells using a wide-bore pipette.
10. In two fresh 50-mL centrifuge tubes add 28-mL of Buffer B. Place these tubes into the high-speed centrifuge.
11. Add 2-mL of your erythrocyte suspension to each of your tubes. The entire class should try to do this as quickly as possible, so wait until everyone is ready.
12. As soon as the last sample is loaded, close the centrifuge, and spin the tubes at 20,000 ×g for 20 minutes.
13. Remove the supernatant without disturbing the "ghost" pellet or the hard-packed button pellet. The ghost pellet contains your plasma membranes, while the hard-packed pellet is unwanted cell debris.
14. Carefully resuspend the "ghost" pellet in 30-mL of Buffer B without disturbing the hard-packed pellet.
15. As soon as the last sample is loaded, close the centrifuge and spin the tubes at 20,000 ×g for 20 minutes.
16. Carefully resuspend the "ghost" pellet in 2-mL of Buffer A without disturbing the hard-packed pellet.
17. Combine re-suspended plasma membrane isolates from both centrifuge tubes into one clean tube, and place on ice.

Assay for Ouabain-Sensitive Na+/K+-ATPase

1. Set up tubes containing 1-mL each of Substrates X and Y. Substrate Y contains ouabain.
2. Bring tubes to 37°C in a water bath. Then add 25-μL of your plasma membrane isolate to each tube.
3. Incubate for 5 minutes.
4. Add 0.5-mL of Buffer C to each tube and transfer to an ice-water bath.
5. Incubate for 6 minutes.
6. Return to the 37°C water bath. Add 1.5-mL of Buffer D to each tube.
7. Incubate for 10 minutes.
8. Transfer 1.5-mL to a microcentrifuge tube, and spin at top speed for 1 minute.
9. Using the spectrophotometer, read the absorbance of the supernatant at 850 nm and record your results for question 1. Give your values to your instructor so that she may calculate and record the class average values.

WORKSHEET

1. Record the A_{850} results for your samples below:

Sample	A_{850}
Substrate A	
Substrate B (ouabain)	

Record the class average values as well.

Sample	Class mean A_{850}
Substrate A	
Substrate B (ouabain)	

2. What is the difference in A_{850} between your two assays? What are you measuring? How does the presence of ouabain effect the A_{850} value? Why did you include this control sample?

___

___

___

___

___

___

___

___

3. How does your sample differ from the class average? Why do you think this is? Discuss the possible factors that may have increased or decreased your values. Keep in mind that your blood samples should be the same for each group.

___

___

___

___

___

___

___

___

4. What would the presence of membranes from other organelles (endoplasmic reticulum, Golgi apparatus, mitochondria) have on your results? Discuss possible variations that may occur in your data if membranes from other membrane-bound organelle were mixed in with your erythrocyte ghosts.

5. What would the presence of other blood cells (lymphocytes, platelets, infection microorganisms) have on your results? Discuss possible variations that may occur in your data if membranes from other cells were mixed in with your erythrocyte ghosts.

DISCUSSION

1. Discuss why red blood cells were chosen for this experiment. What are erythrocyte ghosts?
2. What is the function of Na +/K+-ATPase enzyme in the erythrocyte plasma membrane? Is this protein unique to red blood cells? What is the purpose of the ouabain samples in your assay?
3. Discuss the three types of membrane proteins: integral, peripheral, or lipid-anchored. How are they associated with the plasma membrane, and how is this association built into their function?
4. Discuss the production and processing of a plasma membrane protein. Begin at the nucleus, and end at association with the plasma membrane.
5. Discuss the effect of brefeldin A on a cell. Design an experiment that employs brefeldin A to analyze protein trafficking.

Cell Fractionation

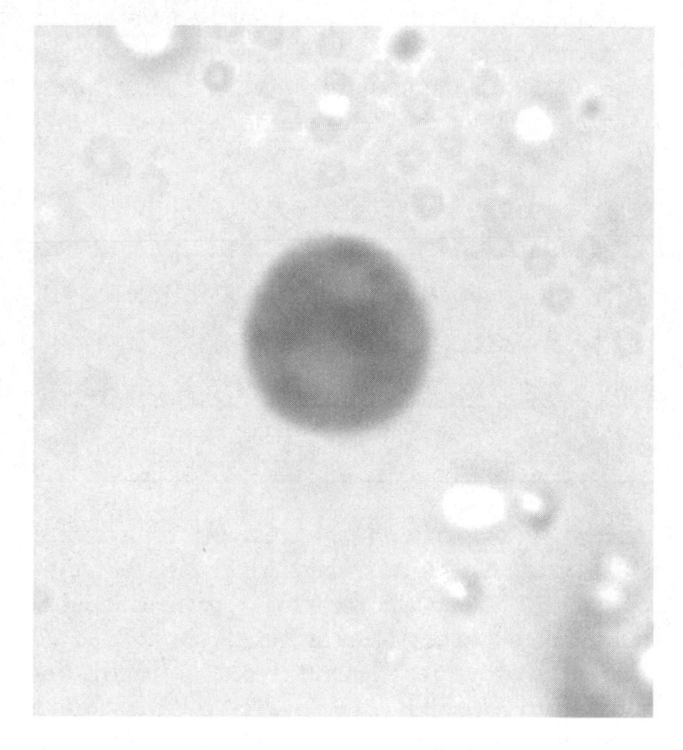

Objectives

- To use differential centrifugation to separate nuclei and mitochondria.
- To stain and observe these "naked" organelle under the microscope.

5

OVERVIEW

DESCRIPTION

The ability to separate and isolate cell components is an essential tool for the cell biologist. Using various centrifugation techniques, one can isolate every part of the cell. These isolates can be used to test the molecular components, the biochemical activity, and identify specific marker proteins or molecules to be used in future experiments. The ability to create "naked" organelles also creates the opportunity to observe these structures without resorting to thin sectioning or freeze fracture techniques, which may disrupt natural morphology. With the right knowledge and an ultracentrifuge, a infinite number of opportunities become available to the budding cell biologist.

CONCEPTS & VOCABULARY

- Differential centrifugation
- Density gradient centrifugation
- Homogenization
- Radius of rotation
- RCF
- RPM
- Sedimentation coefficient

CURRENT APPLICATIONS

- Every membrane-bound organelle can be separated by centrifugation. However, by coupling density-gradient centrifugation with column chromatography, one can create a concentrated sample of one specific organelle.
- Centrifugation can be used to separate just about any solution by density and size. The technique can be used to separate cells of different types, the organelles of the cell, or even different molecules. In fact, in the nuclear sciences, large ultracentrifuges are used to separate different radioactive isotopes, which vary by only a few neutrons.

REFERENCE

Karp, G. (2010). Differential centrifugation. *Cell and Molecular Biology: Concepts and Experiments*. (733–734). Hoboken, NJ: John Wiley & Sons, Inc.

BACKGROUND

CENTRIFUGES AND CENTRIFUGATION

A centrifuge is a device that uses centrifugal force to separate the components of a solution based on variations in density. Centrifugal force is created when rotation is applied to a fixed sample. The rotation applies an outward force to the sample, moving away from the center of rotation. Centrifugal separation in a liquid medium is achieved due to the fact that denser particles in a solution will sediment more rapidly than less dense particles. This concept is called the sedimentation principal. The rate at which a particle travels through a solution is the sedimentation velocity. This value can be used to calculate a sedimentation coefficient, a value specific to the particle being separated. Knowledge of either of these values is very helpful when designing new centrifugation separation protocols.

The centrifuge itself is composed of a rotor with slots for sample tubes or holders. Tubes come in all shapes and sizes, allowing for nanoscale and macroscale samples to be separated. However, the tubes must always fit snugly into the slots of the rotor. The movement of the rotor is driven by a computer-controlled motor that increases or decreases the speed of rotation. The speed of rotation is measured in revolutions per minute (RPM), or the number of full rotations that are achieved in a minute. The relative centrifugal force (RCF) can be determined using the RPM and the radius of rotation of the specific rotor that is being used. RCF is also referred to as g-force or Xg.

There are three general types of centrifuges common to cell biology. The simplest is the clinical or serological centrifuge. These centrifuges are found in most laboratories and hospitals. Typically, they are used to separate solid from liquid: for example, to separate blood cells from plasma. They are the slowest of the centrifuges, capable of speeds up to 14,000 rpms and rcfs of around 20,000 ×g. Typically they only have a single, non-removable rotor and can only hold one or two types of tubes. The next type is the high-speed centrifuge. These centrifuges are faster, and more powerful, capable of speeds up to 20,000 rpms and rcfs of 40,000 ×g. These centrifuges have a variety of removable rotors, capable of handling small (2-mL) tubes or large (500-mL) bottles. Typically, these centrifuges are used to separate solid particles from a liquid, as well. But the high-speed centrifuges allow for larger sample volume and faster sedimentation times. Finally, there are ultracentrifuges, which also have a variety of rotors, which are typically spun under vacuum. They are capable of speeds up to 100,000 rpm and rcfs of 900,000 ×g. Ultracentrifuges are powerful enough to separate out organelle or even molecules within a solution.

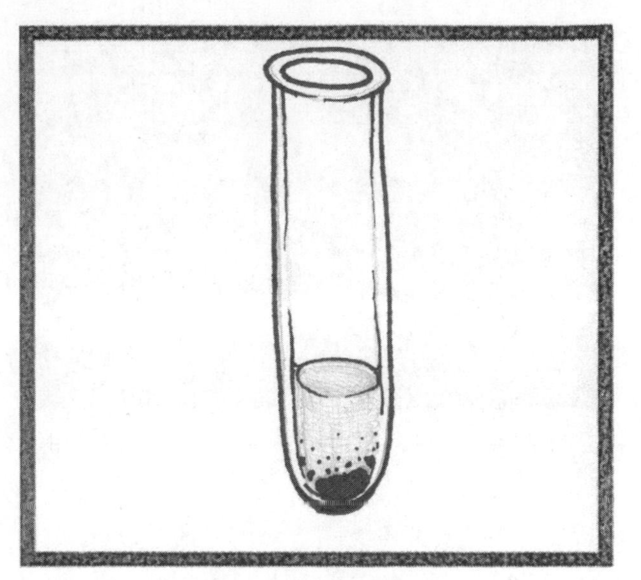

Pellet and Supernatant

High-Speed Centrifuge

TYPES OF CENTRIFUGE ROTORS

There are two main types of centrifuge rotors that you will encounter in cell biology. Fixed-angle rotors are one-piece, solid, metal rotors with slots for tubes or bottles. The sturdy design of these rotors allows for the maximum amount of centrifugal force to be applied to the sample. Ultracentrifuge rotors capable of 900,000 ×g rcfs must be fixed-angle and made of thick, high-grade aluminum. The problem with the fixed-angle rotor is that it shortens the path of the centrifugal force. Typically, sedimented material will be found on the bottom and top side of the tube. Swing-bucket rotors swing outward away from the center of rotation. This allows for the maximum path length for the centrifugal force, and also allows for the formation of flat layers of sediment. However, because these rotors are not composed of a single piece of metal, they are not used for very-high-speed centrifugations.

Centrifuges are extremely useful pieces of laboratory equipment, but due to their powerful, high-speed rotation they are also potentially hazardous. Whenever you are mounting a rotor, be sure to place it into the centrifuge properly. If there is a screw that needs to be tightened to secure the rotor, be sure that it is at least hand-tight. Make sure to balance samples on either side of a rotor by weight. Use a balance to insure that the weight is identical. Check all buckets on a swinging bucket rotor to make sure that they are secure and move freely. When you start your centrifuge run, listen for any strange noises. You might be able to detect an imbalanced rotor before it can do any damage to the machine.

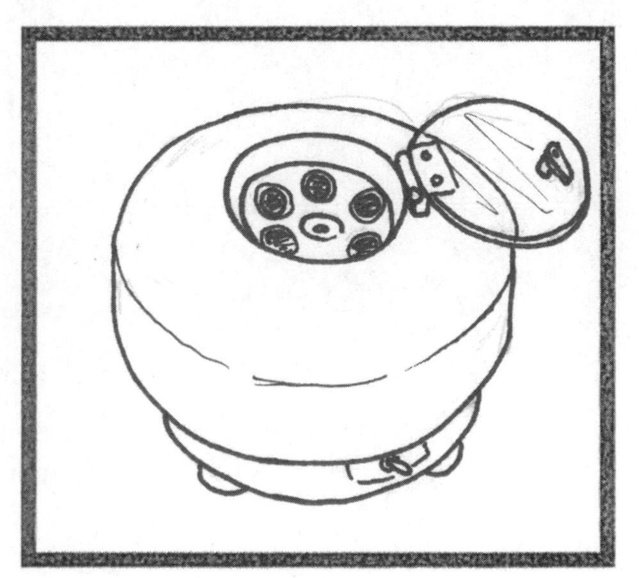

Clinical Centrifuge

A Fixed-Angle Rotor

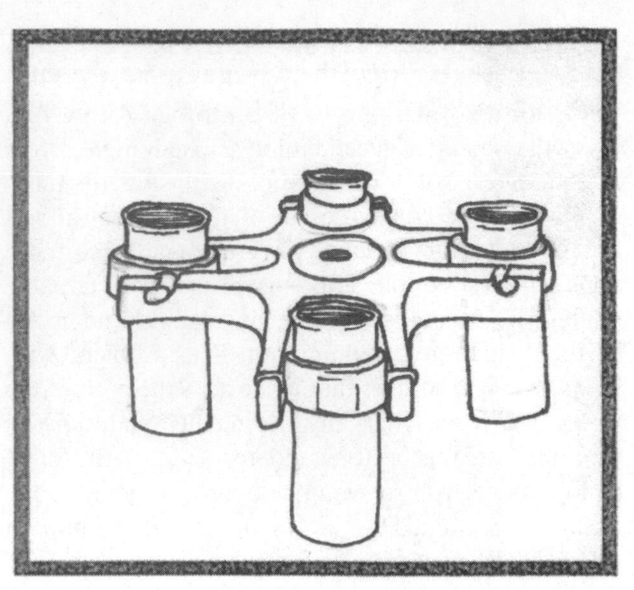

A Swing-Bucket Rotor

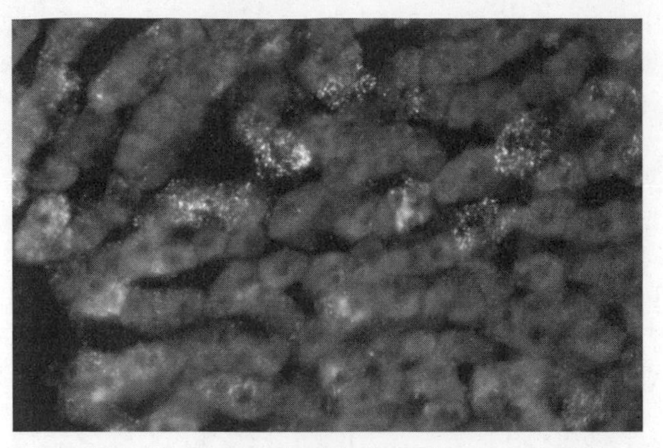

Mitochondria Stained with the Fluorescent Mito-Tracker Dye

DIFFERENTIAL CENTRIFUGATION

In a standard centrifugation, a sample is separated into the solid pellet and the liquid supernatant. With differential centrifugation, several different components can be isolated from a single sample using a series of several centrifugations. After each centrifugation, the pellet is set aside and the supernatant is centrifuged again at a higher speed. For example, today you will be separating nuclei and mitochondria from a sample of homogenized plant tissue. During the first, slower spin, you will pellet out the heavier and denser nuclei. After your first spin, the nuclei will be found in the pellet at the bottom of the tube. Then, using a faster spin, you will pellet out the smaller and less dense mitochondria. This method can be used to separate several different types of organelle; or separate soluble, organelle-bound, and membrane-bound proteins. All you need is a centrifuge capable of low and high speeds (or two centrifuges with different speed ranges).

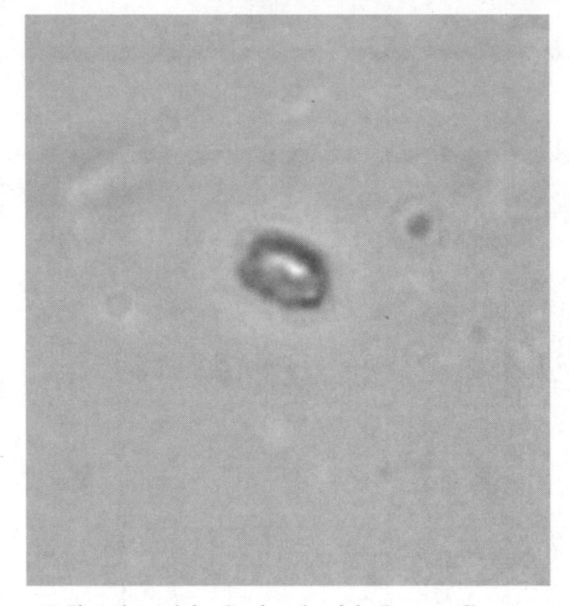

Mitochondria Stained with Janus Green

DENSITY GRADIENT CENTRIFUGATION

If you have access to a swing-bucket rotor, you can perform density gradient separations. The most common form of density gradient uses several solutions of sucrose, each with a decreasing concentration. A gradient is created in a tube by layering each sucrose solution on top of the next, with the densest solution at the bottom and the lightest at the top. This creates several interfaces, called sucrose

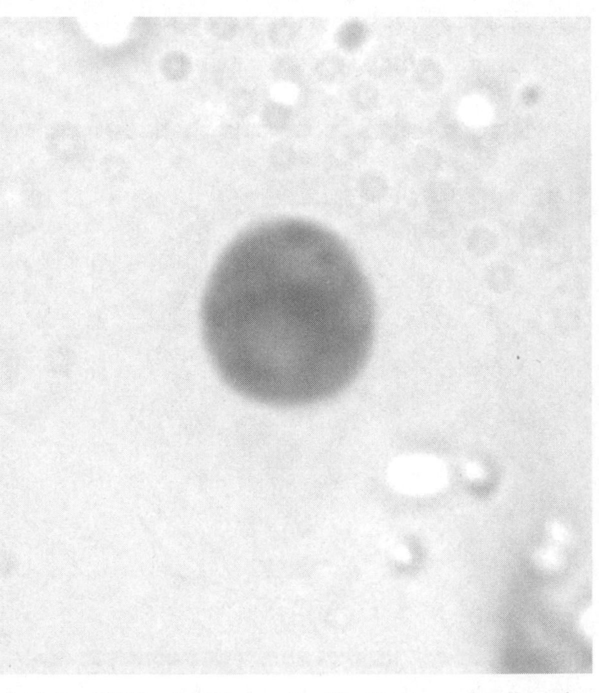

"Naked" Nuclei Stained with Lacto-Acet-Orcein

cushions or pillows. A sample is then layered on the top level, and the entire density gradient is centrifuged. As the components of the sample travel through the gradient, some samples will stop at each interface. This is due to the fact that sedimentation can only occur in a solution that is less dense than the particle. When a particle reaches an interface of higher density, it will stop and accumulate on that sucrose cushion. After the centrifugation, the interfaces can be collected. Specific organelle or molecules will be found at specific interfaces. Other properties, like solubility or hydrophobicity, can also be employed in gradient centrifugation protocols to allow for separation beyond density differences.

PROCEDURES

1. Using a single-edge razor blade, remove a total of 20 g of the outer 2–3 mm of the cauliflower surface.
2. Place the tissue in a chilled mortar with 40-mL of ice-cold mannitol medium and 5 g of sterile, cold, purified sand. Grind the tissue with a chilled pestle for 4 min.
3. Filter the suspension through four layers of cheesecloth into a chilled 50-mL conical centrifuge tube; also wring out the juice into the tube. Allow the filtrate to remain undisturbed for about 2 min, while the sand and debris settle.
4. Carefully decant the supernatant into a fresh, chilled 50-mL conical centrifuge tube, and spin at 600 ×g for 10 min at 0–4°C. Make sure that the centrifuge tubes are balanced.
5. Decant the postnuclear supernatant into a clean, chilled centrifuge tube(s), and place the centrifuge tube with the nuclear pellet in an ice water bath.
6. Centrifuge the postnuclear supernatant at 10,000 ×g for 30 min at 0–4°C. Again, be sure that the centrifuge tubes are balanced. During the centrifugation, the nuclear pellet can be examined microscopically, as described below.
7. Decant and discard the postmitochondrial supernatant, and add 5.0-mL of the ice-cold mannitol medium to the mitochondrial pellet.
8. With a spatula, scrape the mitochondrial pellet from the wall of the centrifuge tube, and then, with a Pasteur pipet, thoroughly resuspend the pellet in mannitol medium.
9. Transfer the mitochondrial suspension to a test tube and place in an ice water bath.

Microscopic Examination of the Nuclear Fractions

1. With a spatula, remove a tiny amount of the nuclear pellet and smear the material on a clean slide. Immediately, before the smear dries, add several drops of the nuclear stain lacto-acet-orcein. After 15 sec, add a cover slip, and very *gently* press out the excess stain with a paper towel.
2. Examine the preparation under high-dry magnification. When you have located a region with nuclei, switch to the oil-immersion objective. For each of five round nuclei, measure the nuclear diameter, and then answer the questions on your datasheet.

Microscopic Examination of the Mitochondria Fractions

1. Place five drops of the mitochondrial suspension in a small test tube. Add five drops of the Janus green stain, swirl the tube, and allow the mixture to remain at room temperature for 10 min.
2. Place a drop of the mixture on a clean slide, add a cover slip, and examine under high-dry and oil-immersion. Identify the mitochondria, which appear as very small, blue-green structures.
3. Measure the length of five mitochondria. Observe the morphology of the mitochondria, and then answer the questions on the datasheet.

WORKSHEET

1. Measure the radius of rotation for one of the centrifuges in the lab. Convert the following rpm values into RCF values: 100 rpm, 350 rpm, 770 rpm, 2000 rpm.

2. When would you use a swing-bucket rotor? When would you use a fixed-angle rotor? What are the advantages of each kind of centrifugation?

3. Draw an image of a naked mitochondrion below. Be sure to label any important structures, mark the magnification and add a scale bar.

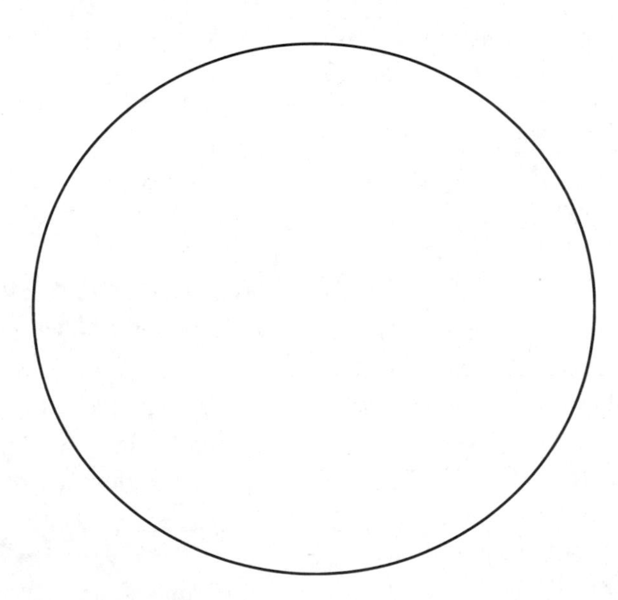

4. Describe the difference between your observations of the naked nucleus versus your previous observations of nuclei within cells.

5. Using Figure 18.35 in Karp, draw a labeled figure of a sucrose density gradient centrifugation sample with 10% sucrose, 25% sucrose, 50% sucrose. Indicate at which interface you would expect the nuclei to settle from a plant tissue sample that passes through the gradient. Indicate at which interface you would expect the chloroplasts to settle. Indicate at which interface you would expect the mitochondria to settle.

DISCUSSION

1. Define differential centrifugation. Define density gradient centrifugation. Discuss the differences between these two extraction methods.
2. Describe the appearance of your isolated nuclei. How does their appearance differ from the nuclei you have seen in previous experiments?
3. Describe the appearance of your isolated mitochondria. How does their appearance differ from the mitochondria you have seen in your textbooks or the literature?
4. Today, you isolated nuclei and mitochondria from cauliflower florets. Describe an experimental design in which you could isolate chloroplasts. Which organism and tissue would you choose? How would you set up the fractionation?
5. What are the limitations of fractionation and separation by centrifugation? For example, what would happen if you wanted to separate two different types of vesicles that were the same size? What if you wanted to separate two different types of vesicles that had the same membrane consistency? Discuss what you could do to identify one vesicle from the other.

Isolation and Activity of Mitochondria

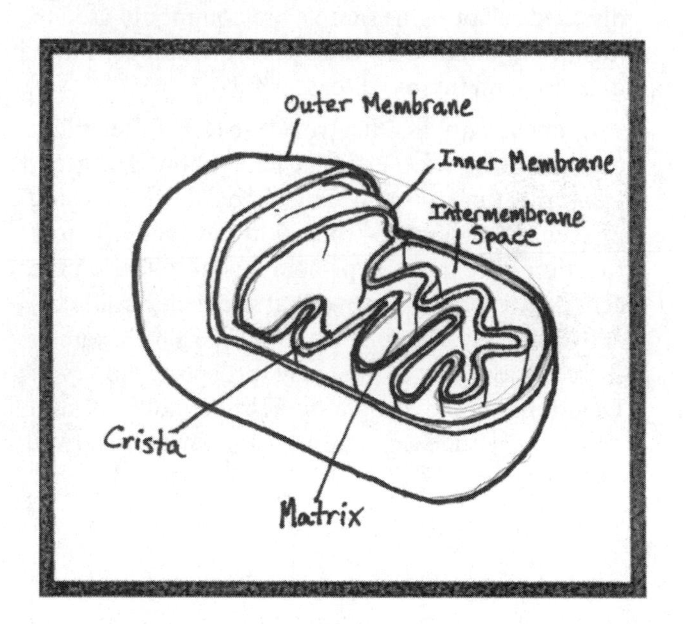

Objectives

- Isolate naked mitochondria from cauliflower.
- Determine the succinate dehydrogenase activity of your isolate.

DESCRIPTION

The mitochondrion is the energy plant of the cell. It is a distinct organelle, wrapped in two membranes. This structure allows for the creation of two distinct spaces: the matrix and the inter-membrane space. Fingerlike-structures called cristae, composed of projections of the inner membrane, increase the surface area of the mitochondrial interior. The components of the TCA Cycle, the Electron Transport Chain, and fatty-acid catabolism are all found within the mitochondrion. The structure of the mitochondria allow for the creation of two distinct environments, separate and safe from the cytoplasm. ATP synthesis can produce harmful by-products, like reactive oxygen species, which are contained with the two membranes of the organelle.

CONCEPTS & VOCABULARY

- ATP synthase
- Cristae
- Glycolysis
- Electron transport chain
- Inter-membrane space
- Mitochondrial matrix
- Oxidative phosphorylation
- Reactive oxygen species
- TCA cycle

CURRENT APPLICATIONS

- Isolation of intact organelles is a key skill for the cell biologist. All organelle-specific proteomic and activity assays begin with the removal of all other cell components. For most model organisms, the common complement of proteins for each organelle is already determined. Today, cell biologists are using this knowledge to determine the protein components of disease. Differences in the protein profile of a diseased cell or tissue sample are commonly used to identify a number of disorders, including cancer, diabetes, and thyroid disease.
- Enzyme activity assays give a researcher the ability to see the true effects of gene expression. The gene, transcript, and protein can all be detected using molecular techniques, but without activity and function the expressed protein has no effect on the cell. Activity assays prove that the gene product is active and functioning within the cell. Designing these assays is not an easy prospect, however. Determining the proper *in vitro* conditions and detection method can take months of trial and error.

REFERENCE

Karp, G. (2010). Metabolism; aerobic respiration and the mitochondrion. *Cell and Molecular Biology: Concepts and Experiments*. (105–113; 173–200). Hoboken, NJ: John Wiley & Sons, Inc.

BACKGROUND

FUNCTIONS OF THE MITOCHONDRIA

The majority of energy in the cell is produced by the mitochondria. While glycolysis and fermentation occurs in the cytoplasm, the tricarboxylic acid cycle (TCA cycle) and oxidative phosphorylation via the electron transport chain both occur in the mitochondria. The TCA cycle and the electron transport chain produce over 30 ATPs for every glucose molecule metabolized. Pyruvate from glycolysis is converted into Acetyl-CoA as it is shuttled into the mitochondria. Fatty acid catabolism also occurs in the mitochondria. From each fatty acid, a Fatty Acyl-CoA is broken off a carbon chain and shuttled off into the mitochondria. This Fatty Acyl-CoA is converted into one acetyl-CoA where it enters the TCA cycle, producing 1-NADH and 1-$FADH_2$ (equivalent to 4 ATP per carbon removed from the fatty acid).

STRUCTURE OF THE MITOCHONDRIA

The mitochondria are surrounded by the double membrane system, composed of an outer membrane and an inner membrane. Between these two membranes is a region called the inter-membrane space. The space beyond the inner membrane and the core of the mitochondria is called the matrix. Long finger-like structures are created from projections of the inner membrane pushing into the matrix. These structures are called cristae.

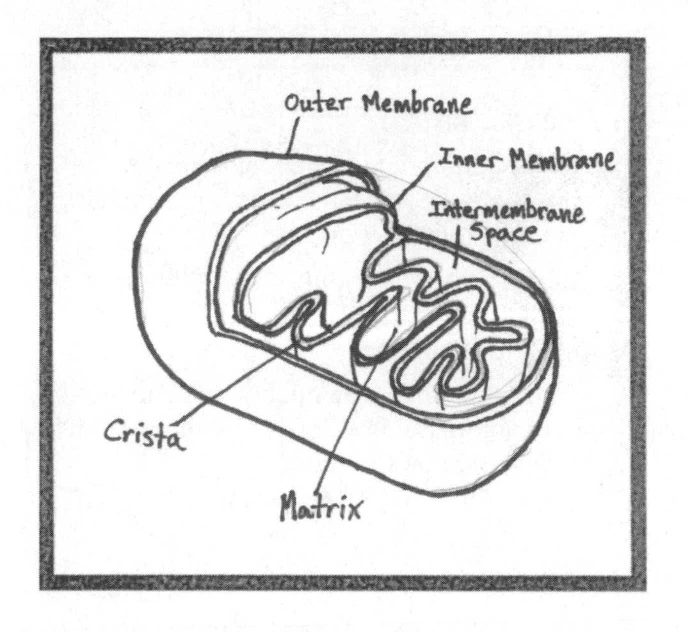

Both the electron transport chain and fatty acid chain (β-fatty acid oxidation) have the potential to produce reactive oxygen species (ROS). Enzymes within the mitochondria are used to remove these harmful by-products. However the inclusive structure of the mitochondria creates a barrier of protection for the rest of the cell. The compartmentalization of the inter-membrane space and the cristae allows for the creation of two distinct environments. For example, the aqueous environment of the inter-membrane space is maintained at a very low pH, which allows for the production of ATP via the proton pump of ATP-synthase. The electron transport chain components embedded in the cristae membrane release H+ ions into the inter membrane space through ATP-synthase. This proton pump provides energy for the production of ATP. The entire TCA cycle is localized to the matrix.

SUCCINATE DEHYDROGENASE ACTIVITY

Today you will be isolating mitochondria and observing the activity of one of the enzymes found in the matrix. Succinate dehydrogenase is an enzyme that converts succinate into fumarate within the TCA cycle. This reaction also reduces FAD+ into $FADH_2$, which is later used to provide electrons for the electron transport chain. You will perform an assay to measure the succinate dehydrogenase within the sample of naked, intact mitochondria. To accomplish this, you will use an artificial electron acceptor, 2,6 dichlorophenolindophenol (DCIP) along with the addition of sodium azide, which will inhibit the transfer of electrons from cytochrome a to oxygen. The electrons will reduce the blue DCIP and turn it into a colorless product. You will the use a spectrophotometer to measure the absorbance at 600 nm at 0 minutes and 5 minutes to measure the progress of the reaction. Since succinate will be the only TCA cycle reactant added to the assay, only succinate dehydrogenase will be active, still producing fumarate and $FADH_2$.

ENZYME KINETICS

To revisit enzyme kinetics, remember that the initial events of enzyme catalysis are summed up by the Michaelis-Menten equation. They theorized that the enzyme (E) and substrate (S) combine reversibly to form the Enzyme-Subtrate complex (ES). The ES then dissociates into free enzyme and Product (P). In the reaction below, k_1 is the forward rate constant, k_2 the reverse.

$$E + S \underset{k_2}{\overset{k_1}{\leftrightarrow}} ES \overset{k_3}{\rightarrow} E + P$$

The initial velocity of the reaction rate can be calculated using the rate constants.

$$v_0 = \frac{k_3[E][S]}{(k_2 + k_3/k_1) + [S]}$$

Since the rate constants are constant, and the substrate concentrate is known, the initial velocity is directly proportional to the enzyme concentration. You will determine the initial velocity of three different concentrations of mitochondrial isolate.

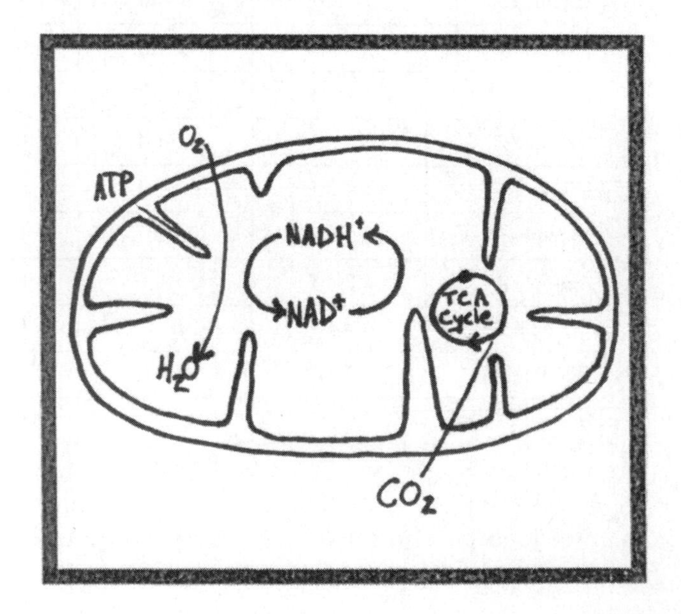

PROCEDURES

Isolation of Mitochondria

1. Follow the procedures in the Cell Fractionation experiment, saving the mitochondrial pellet and adding 7.0-mL of mannitol assay medium to resuspend the pellet.

2. Keep this suspension on ice. Be sure to gently mix the suspension each time before you draw fluid from the tube.

Measurement of Succinate Dehydrogenase Activity

1. Turn on your spectrophotometer and set it for absorbance at 600 nm.
2. Label 10 cuvettes Blanks 1, 2, and 3, and Samples 1 through 7. Remember to keep all reaction mixtures at room temperature, but keep your mitochondrial suspension on ice.
3. Make Sample 7 first and incubate it in a boiling water bath for 5 minutes, then put it on ice.
4. To make each of the other 9 blanks/samples, add the components in the following order: assay medium, azide, DCIP, malonate (for Sample 4), and succinate (except for Sample 9). Cover with Parafilm and invert 4-times to mix.
5. Move to the spectrophotometer.
6. Adjust for Blank 1.
7. Add the appropriate amount of mitochondrial suspension to Sample 1, cover with Parafilm and invert 4-times to mix.
8. Measure the sample for the 0-minute data point immediately.
9. Adjust for Blank 2.
10. Add the appropriate amount of mitochondrial suspension to Sample 2, cover with Parafilm, and invert 4 times to mix.
11. Measure the sample for the 0-minute data point immediately.
12. Adjust for Blank 3.
13. Add the appropriate amount of mitochondrial suspension to Sample 3, cover with Parafilm, and invert 4 times to mix.
14. Measure the sample for the 0-minute data point immediately.
15. Repeat steps 13 and 14 for Samples 4–7.
16. Set the cuvettes aside at room temperature.
17. Measure all of the samples every 5 minutes for another 20 minutes. Remember to adjust for Blank 1, then measure Sample 1; adjust for Blank 2, then measure Sample 2; and adjust for Blank 3, then measure Samples 3–7.
18. Collect your data on the datasheet.

Tube	Assay Medium	Azide	DCIP	Malonate	Succinate	Mitochondrial Suspension
Blank 1	3.7-mL	0.5-mL			0.5-mL	0.3-mL
Sample 1	3.2-mL	0.5-mL	0.5-mL		0.5-mL	0.3-mL
Blank 2	3.1-mL	0.5-mL			0.5-mL	0.9-mL
Sample 2	2.6-mL	0.5-mL	0.5-mL		0.5-mL	0.9-mL
Blank 3	3.4-mL	0.5-mL			0.5-mL	0.6-mL
Sample 3	2.9-mL	0.5-mL	0.5-mL		0.5-mL	0.6-mL
Sample 4	2.7-mL		0.5-mL	0.2-mL	0.5-mL	0.6-mL
Sample 5	3.4-mL		0.5-mL		0.5-mL	0.6-mL
Sample 6	3.4-mL	0.5-mL	0.5-mL			0.6-mL
Sample 7	2.9-mL	0.5-mL	0.5-mL		0.5-mL	0.6-mL

WORKSHEET

1. Write down the absorbance for each tube in the space below

Tube	0-minutes	5-minutes	10-minutes	15-minutes	20-minutes
Sample 1					
Sample 2					
Sample 3					
Sample 4					
Sample 5					
Sample 6					
Sample 7					

Now calculate the change in absorbance between each time point between 0 and 20-minutes.

Tube	ΔAbs 0–5 m	ΔAbs 5–10 m	ΔAbs 10–15 m	ΔAbs 15–20 m
Sample 1				
Sample 2				
Sample 3				
Sample 4				
Sample 5				
Sample 6				
Sample 7				

On the graph paper provided, plot the ΔAbs versus time. Use 0 as a ΔAbs for 0-minutes. Draw a best-fit curve with a ruler, and label each line with the Sample number.

2. Estimate the initial velocity by using the change in Absorbance for each sample between 0 and 5-minutes. Which volume of Sample gives the highest initial velocity? Explain why.

____

____

____

____

____

3. For Samples 1–3, how does the change in the amount of suspension change the initial velocity of the reaction? Is this a linear change: as the amount doubles, the rate doubles? Explain your answer.

____

____

____

____

____

4. Sample 4 contains malonate, a competitive inhibitor of succinate dehydrogenase. How does the initial velocity of Sample 4 differ from Sample 3? Why is this the case?

5. Sample 5 does not contain sodium azide. Sodium azide prevents the electron transport chain from completing. How does the initial velocity of Sample 5 differ from Sample 3? Why is this the case?

6. Sample 6 does not contain succinate, the substrate of succinate dehydrogenase. How does the initial velocity of Sample 6 differ from Sample 3? Why is this the case?

7. Sample 7 contains a heat-treated mitochondrial isolate. How does the initial velocity of Sample 7 differ from Sample 3? Why is this the case?

DISCUSSION

1. Discuss why succinate dehydrogenase was chosen for this assay, instead of another enzyme from the TCA cycle.
2. Discuss why sodium azide was included in the assay. How would the removal of this compound change the data collected?
3. Malonate is a competitive inhibitor for succinate dehydrogenase. How would the addition of this compound change the data collected?
4. What would heat treatment in boiling water do to the proteins in the mitochondria isolate? How would this treatment change the data collected?
5. The succinate dehydrogenase in your samples will consume the available succinate as the reaction progresses. The activity of succinate dehydrogenase will slowly decrease in a naked mitochondrial isolate at room temperature. How will these two factors change the data collected over a 30-minute time course?

Staining of the Extracellular Space

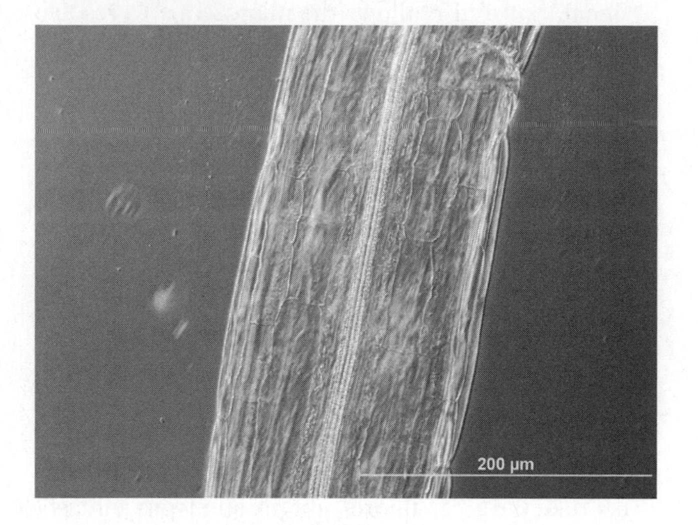

Objectives

- Observe prepared slides that highlight plant plasmodesmata.
- Observe fluorescently labeled slides that highlight extracellular matrix proteins.

DESCRIPTION

Multicellular organisms are composed of tissues and organs, and very rarely is one cell more important than the thousands of cells that surround it. Most tissues are composed of one or more types of cells embedded in a collection of biomolecules loosely referred to as the extracellular matrix (ECM). Epithelial tissue is composed of several tight layers of cells, all in contact with each other. Connective tissue, on the other hand, consists of one or more cells that are scattered throughout the tissue. The extracellular matrix is an organized network of proteins near the exterior of the plasma membrane. The ECM may be composed of collagen, fibronectin, laminin, and/or proteoglycans. The ECM provides mechanical support and releases signals and compounds that are essential for cell survival, allows passage through the substratum for motile cell migration, and is important for tissue separation or a molecular barrier. While animal cells rely on the ECM for their support, bacteria, fungi, and plants maintain an additional structure called the cell wall. Cells within a tissue may be connected to each other in a variety of ways. Gap junctions in epithelia cells and plasmodesmata in plant cells are direct connections between cells, characterized by a contiguous channel of cytoplasm. Tight junctions are cell connections where only the plasma membranes are in contact. Today, you will stain and observe the epithelia and connective tissues so that you may observe and sketch them. You will also observe the apoplast and plasmodesmata of plant tissues.

CONCEPTS & VOCABULARY

- Adherins
- Caherins
- Collagen
- Desomsomes
- Fibronectin
- Gap junctions
- Laminin
- Immunoglobins
- Integrins
- Plasmodesmata
- Proteoglycans
- Selectins
- Tight junctions

CURRENT APPLICATIONS

- Confocal microscopy uses a laser to excite fluorescent molecules in a histological sample. Since the laser can excite the molecule at high energy and is limited to a smaller wavelength range, it can cause a chromophore to emit a very strong signal. This signal is detected by the computer for a specific wavelength range, removing extraneous signal. This not only produces better resolution of signal, but it also allows the microscope to resolve a thinner focal plane. In some cases, the focal plane of a confocal microscope is 10 to 100-times thinner than a UV-Fluorescence Light Microscope.
- Fluorescent labels (chromophores) come in a variety of colors. Each of these colors has a specific range of excitation and emission wavelengths. Chromophores whose excitation/emission ranges do not overlap can be used together in the same sample. These molecules can be used to label proteins, antibodies, drug molecules, and many other biomolecules. The genetic coding sequence for some fluorescent proteins can be added to the end of recombinant proteins to directly label a gene product (like the green fluorescent protein from jellyfish, or GFP).

REFERENCE

Karp, G. (2010). Interactions between cells and their environment. *Cell and Molecular Biology: Concepts and Experiments.* (230–263). Hoboken, NJ: John Wiley & Sons, Inc.

BACKGROUND

TISSUE STRUCTURE

It is easy to discuss cell biology in terms of unicellular organisms. By studying yeast, bacteria, or protists, we can focus on everything from the plasma membrane inward. However, if we are to focus on complex, multicellular organisms, the extracellular space becomes more and more important. Multicellular organisms are composed of tissues and organs, and very rarely is one cell more important than the thousands of cells that surround it. Most tissues are composed of one or more types of cells embedded in a collection of biomolecules loosely referred to as the extracellular matrix (ECM). Each cell in a type of tissue can interact with its neighbors as well as the ECM itself.

To discuss the basic nature of the extracellular space, let's look at two mammalian tissue types: epithelial tissue and connective tissue. Epithelial tissue is composed of several tight layers of cells, all in contact with each other. Beneath these layers, separating the epithelial tissues from all the other tissues is the basement membrane. The basement membrane is a thin layer of extracellular proteins that act as a substrate for the epithelia tissue. Cell-to-cell communication and cell-to-substratum communication direct the growth, activity, and function of all cells in the epithelial tissue. Connective tissue, in contrast, consists of one or more cells that are scattered throughout the tissue. These cells, primarily fibroblasts, are embedded in extracellular material composed of various reticular, collagen, and elastin fibers. Instead of relying on cell-to-cell contact, or cell-to-substratum contact, cells in connective tissues rely on cell surface receptors, called integrins. Integrins allow for communication and interaction, primarily with the extracellular material.

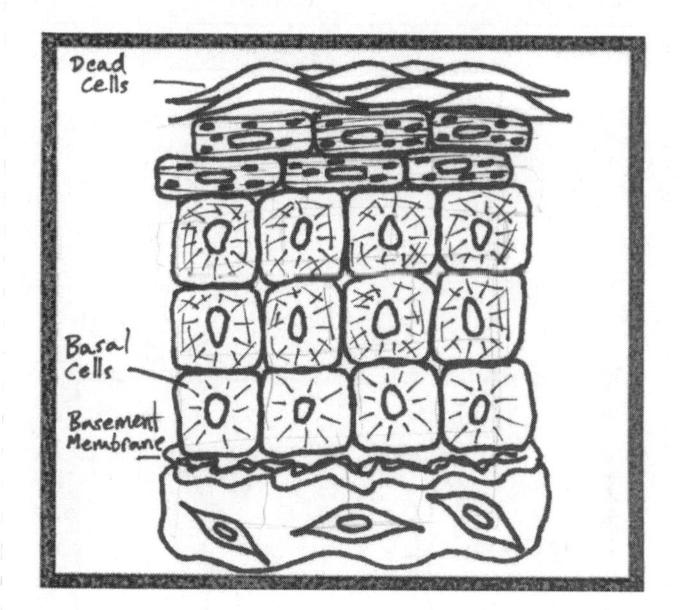

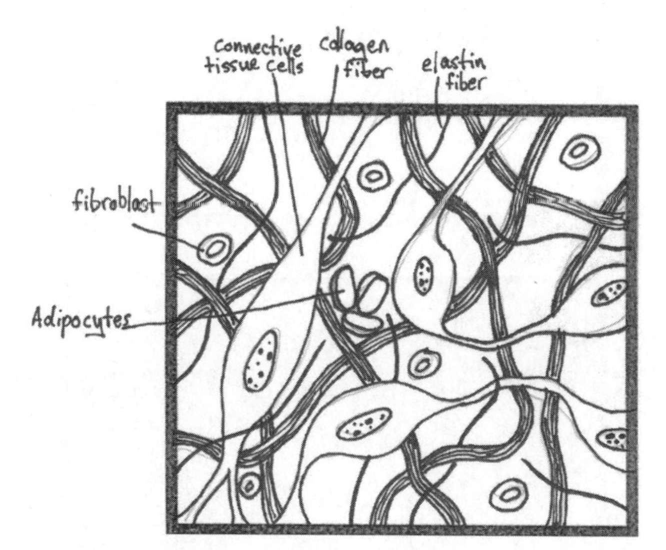

THE GLYCOCALYX & THE EXTRACELLULAR MATRIX

The glycocalyx ("cell") is a halo of carbohydrate molecules bound to plasma membrane, proteins, and lipids. On some cells, the glycocalyx is not very prevalent, while on others (like the cells that line the digestive tract) it is dense and obvious under the microscope. Glycocalyx is important for the mediation of cell-to-cell and cell-to-substratum interactions. It also offers mechanical protection, a barrier to particles, and may be involved in regulation at the cell surface.

The extracellular matrix (ECM) is an organized network of proteins near the exterior of the plasma membrane. The ECM may be composed of collagen, fibronectin, laminin, and/or proteoglycans. These fibrous proteins act as packing material and extracellular adhesive, or they may perform a regulatory role in the shape and function of the cell/tissue. A good example of a common extracellular matrix is the basement membrane of the epidermis. Other tissues that have a prevalent basement membrane ECM include nerve fibers, muscle, and adipose tissue, the lining of the digestive and respiratory tract, and the inner lining of blood vessels. The ECM provides mechanical support and releases signals and compounds essential for cell survival; it allows passage through the substratum for motile cell migration, and it is important for tissue separation or a molecular barrier.

CELL WALLS AND THE APOPLAST

While animal cells rely on the ECM for their support, bacteria, fungi, and plants maintain an additional structure called the cell wall. The primary structure of the cell wall is typically carbohydrate, occasionally combined with proteinaceous molecules. The cell wall provides structure and support, it helps to protect against damage and infection, and it may be involved in the mediation of the cell-to-cell interactions. In plant cells, the primary cell wall is composed of polysaccharides: hemicelluloses and pectin. Various proteins, like expansins, which are necessary for cell wall reorganization and growth, are also found in the cell wall. In woody plant species, a stronger secondary cell wall may be formed that incorporates a stiff polysaccharide called lignin. The space just beyond the exterior of the plasma membrane is called the apoplast. The apoplast includes the cell wall, extracellular fluid, air spaces, and any neighboring cell walls. The apoplast is involved in cell wall construction, the release of hormones and other signals, and the movement and storage of water and solutes.

CELL-TO-CELL INTERACTIONS

Cells within a tissue may be connected to each other in a variety of ways. Gap junctions in epithelia cells and plasmodesmata in plant cells are direct connections between cells, characterized by a contiguous channel of cytoplasm. This allows for diffusion between the two cells and direct signaling. Tight junctions are cell connections where only the plasma membranes are in contact. Integrins on the surface of cells mediate interactions between the cells and the substratum of the tissue. Binding of an internal ligand with an integrin causes inside-out signaling. Binding of an external ligand to an integrin causes outside-in signaling. It is through the integrins that the cells interact with the tissues and vice versa.

Today you will stain and observe the epithelia and connective tissues so that you may observe and sketch them. You will also observe the apoplast and plasmodesmata of plant tissues.

ECM proteins	Adhesion molecules
Collagen	Selectins
Fibronectin	Integrins
Laminin	Cadherens
Proteoglycan	Immunoglobins

PROCEDURES

Observation of Plant Tissue and Prepared Fluorescent-Labeled Extracellular Matrix Slides

1. Obtain a persimmon cross-section slide from your instructor. Observe the slide using the light microscope and the dry/oil objectives. Try to find a good example of a plasmodesma.
2. Sketch your observations for the slide for question 1.
3. Obtain one of three differentially labeled ECM slides from your instructor. This slide will have four fluorescently labeled signals (see the table below).
4. Observe the slide using a UV-fluorescence microscope and dry/oil objectives (check with your instructor for which filter cubes to use).
5. Sketch your observations for each slide for question 2.
6. If available, take a digital microphotograph of each cytoskeletal element and create a merged image (see Appendix A).

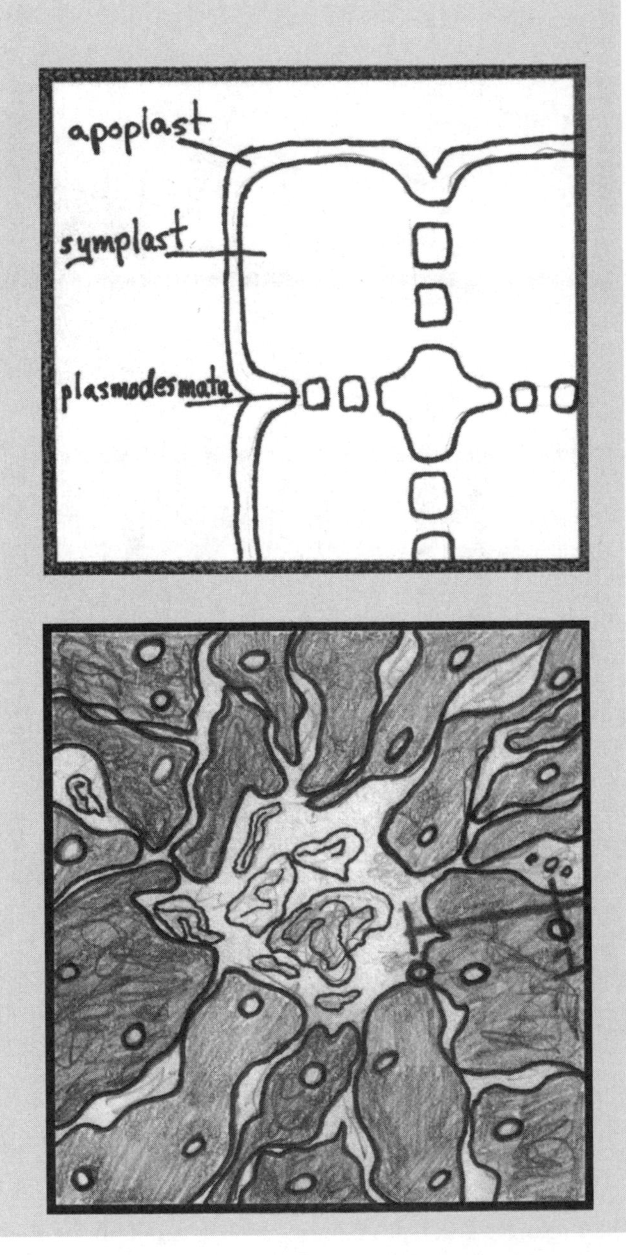

Slide	DAPI-filter	FITC-filter	TRITC-filter	Cy5-filter
Skin	Nucleus	Actin	Elastin*	Collagen*
Gut	Nucleus	Tubulin	Laminin*	Actin
Pancreas	Nucleus	Actin	Insulin	Laminin*

*Represents extracellular matrix components.

7 OVERVIEW

WORKSHEET

1. Observe your prepared persimmon cross-section slides. Can you find plasmodesmata? Draw an example sketch in the space provided below. Remember to label all microscope images with the following: organism, cell type, magnification, and stain; add a scale bar at the bottom right of each sketch.

2. Are all of the cells connected by plasmodesmata? Are there multiple plasmodesmata in every cell? Are there any damaged/dead cells in your cross-section? Describe those cells and the cells that neighbor them.

____

____

____

____

____

____

3. Observe your three pre-stained extracellular matrix slides using epifluorescence. Which ECM proteins are present in each slide? How do they form the structure of the matrix? Draw an example sketch for each slide in the space provided below. Label each of the components that have been labeled with fluorescence. Remember to label all microscope images with the following: tissue type, magnification and stains; add a scale bar at the bottom right of each sketch.

7

OVERVIEW

4. Choose one of the three slides that you observed. Which ECM protein(s) were labeled? Describe where they were found in your sample in comparison to the other structures labeled in the sample. Define the physical properties and functions of this protein(s) and explain how they help to form the structure of the tissue.

5. Consider the gut and pancreas slides that your observed. Which ECM proteins were labeled in these slides? Compare and contrast the placement of the proteins in these samples. How did the proteins add to the structure of the tissue?

DISCUSSION

1. Discuss the physical properties of collagen, elastin, laminin, and proteoglycans. Explain how these properties are used in the construction of the extracellular matrix.
2. Plasmodesmata and gap junctions are direct connections of cytoplasm in between neighboring cells. Discuss the benefits of this type of cell connection. Discuss the problems involved with this type of cell connection.
3. Discuss the methods of cell-to-cell communication found in epithelial tissue. Discuss the methods of cell-to-cell communication found in connective tissue.
4. Discuss the composition, properties, and purpose of the plant cell wall.
5. The extracellular matrix serves other functions besides structural support. Discuss these functions in terms of epithelial and connective tissues.

OVERVIEW

8 Protoplast Isolation and Vacuole Staining

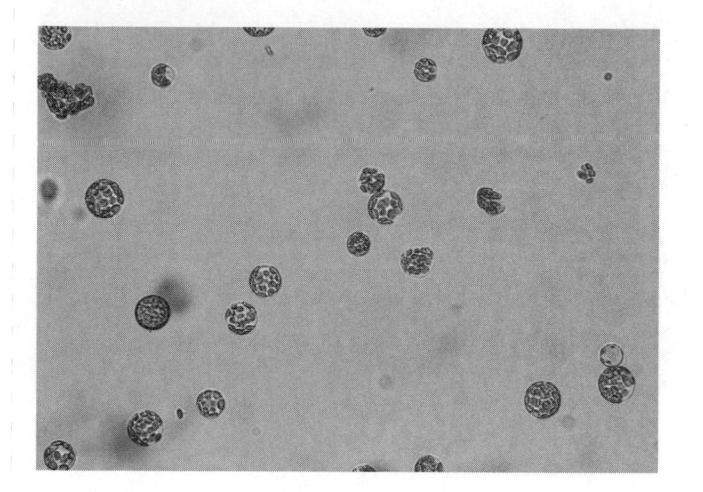

Objectives

- Isolate protoplasts from spinach leaves.
- Observe the protoplasts using light microscopy.
- Stain and observe the central vacuole.

DESCRIPTION

Plant cells are divided in the extracellular space (apoplast) and the intracellular space (protoplast), which are separated by the plasma membrane. The plant cell is protected by a hard, polar cell wall composed of cellulose, hemicellulose, and pectin. The cell wall is semi-permeable, typically allowing smaller molecules through and keeping larger molecules away from the plasma membrane. Unfortunately, many of the common histological stains will not make it through the cell wall or into the protoplast. However, using fungal enzymes (cellulase and macerozyme) designed to break down cell walls, protoplasts can be released from the surrounding tissues and isolated into a hypertonic buffer. These *naked* protoplasts can now be used for *in vivo* physiological or histological experiments. In this experiment, you will learn the basic protoplast isolation technique using spinach leaves. After you have isolated and observed your protoplasts, you will stain them with the acidophilic stain: Neutral Red. Neutral Red is transported through the plasma membrane and through the vacuolar membrane (tonoplast) where it concentrates in the vacuolar lumen. This will allow you to observe the vacuole within the protoplast cells.

CONCEPTS & VOCABULARY

- Acidophilic stain
- Apoplast
- Cellulase
- Lumen
- Macerozyme
- Neutral Red
- Protoplast
- Tonoplast
- Vacuole

CURRENT APPLICATIONS

- Protoplasts can be isolated from any plant tissue. These cells will retain some of the original properties of their source tissue, but much cellular identity comes from interactions with surrounding cells.
- Protoplasts can be made from solid callus cultures and liquid suspension cell cultures. A lower concentration of enzymes is needed to lossen the cell wall material, but the procedure is essentially the same.
- Properly prepared protoplasts can be transformed with expression vectors using chemical or electroporation transformation. This allows for the expression of novel proteins in these cells, for *in vivo* studies.

REFERENCES

Karp, G. (2010). Plant vacuoles; cell culture. *Cell and Molecular Biology: Concepts and Experiments*. (301; 731–733). Hoboken, NJ: John Wiley & Sons, Inc.

Nishimura, M, Akazawa, T. (1975). Photosynthetic activities of spinach leaf protoplasts. *Plant Physiology*. *55*, 712–716.

BACKGROUND

PLANT CELL CULTURE

Cell cultures allow researchers to study physiological responses at a cellular level. Multicellular organisms are composed of various tissues, each composed of one or more different cell types. Mammalian cell cultures maintain some of the identity of their source tissue. Plant cell cultures, on the other hand, quickly lose any identity of their source tissue and begin to resemble undifferentiated cells. Plant cell cultures are produced by taking tissue (explant) from a healthy plant and growing it in a solid medium containing a specific ratio of two plant hormones (an auxin and a cytokinin). The tissue will de-differentiate into a mass of cells called a callus. Callus cells lack chlorophyll and photosynthesis, and require supplemental nutrition to grow and divide. Callus will spread across the surface of the plate, and must be subcultured to fresh medium every few weeks to maintain a healthy culture. Callus can also be placed into liquid medium and, with proper agitation and subculturing, will become a suspension cell culture composed of individual cells. Both callus and suspension cell cultures can be used to study general cellular responses in a species, but they cannot be used to look at tissue-specific responses. Whole organisms must be employed in any tissue-specific experiment in plants. One more unique aspect of plant tissue culture is that callus can be transferred to medium containing a higher concentration of cytokinin, and will form shoots and roots. These plantlets will grow into a new plant, genetically identical to the source tissue. This technique is used quite often as a method of plant propagation.

PLANT PROTOPLASTS

The plant cell is divided into two parts: the apoplast and the protoplast. The apoplast is the portion of the cell external to the plasma membrane, and the protoplast is the portion internal to the plasma membrane. The apoplast is composed of the cell wall and the extracellular space. The protoplast is composed of the nucleus, the cytoplasm, and its contents, including the vacuole, which is surrounded by another membrane called the tonoplast.

The protoplast and all of its contents can be isolated by digestion of the cell wall. Digestion and maceration of leaves and other tissues will cause the release of a number of protoplasts. These protoplasts are short-lived, but may be used to test the

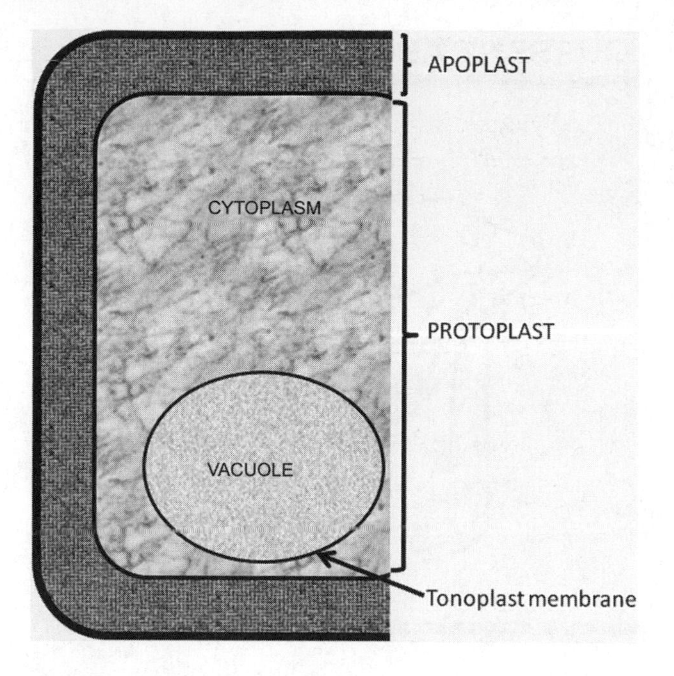

physiology of individual plant cells. Protoplasts maintain the genetic identity of the source tissue, but can also be transformed using plant expression vectors carrying a foreign transgene. Transient expression in protoplasts can be used to study short-term effects of novel gene expression, or intracellular localization of a gene product fused to a fluorescent marker.

Without a cell wall, protoplasts must be kept in hyperosmotic medium, or growth medium containing a much higher amount of a physiologically inactive solute. For many protoplast media, the sugar alcohols mannitol or sorbitol are employed. While plant cells can absorb and metabolize sugar alcohols, they do so at a very slow rate. Given the limited life span of protoplasts (typically 96-hours or less), these accessible and inexpensive solutes are ideal.

PROCEDURES

Spinach Protoplast Isolation

1. Slice leaves to create 0.5–1.9-mm strips. After slicing, submerge leaves into a Petri-dish containing the fresh enzyme solution.
2. Cover the Petri dish and wrap in aluminum foil. Incubate in enzyme solution for 2-hrs at room-temperature with orbital shaking (at ~40 rpm).
3. Sieve the solution through 70 nm nylon mesh funneled into a small beaker (four layers of cheesecloth is an acceptable substitute, but the nylon mesh will allow for more uniform cell size).
4. Pellet cells in a round-bottom tube at 500 ×g (setting 2 on a clinical centrifuge) for 5-minutes at 10°C.
5. Remove the enzyme solution. Add Mannitol/MES buffer, gently re-suspend cells by inversion, and incubate on ice for 5-minutes.
6. Centrifuge again to pellet cells; remove all buffer.
7. Gently re-suspend in 6-mL of W5 Buffer.
8. Observe cells on the slide without a coverslip, or in a welled-slide with a coverslip. Use a low, dry-objective. If an inverted microscope is available, protoplasts may be transferred to 6-well titer plates and observed in the well. **Always transfer protoplasts using wide-bore-tipped pipettes.** These can be made by cutting off the first 1–2 cm of a pipette tip.

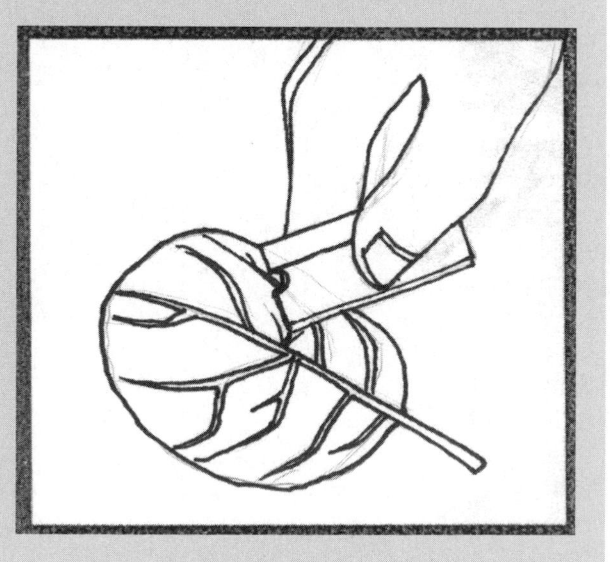

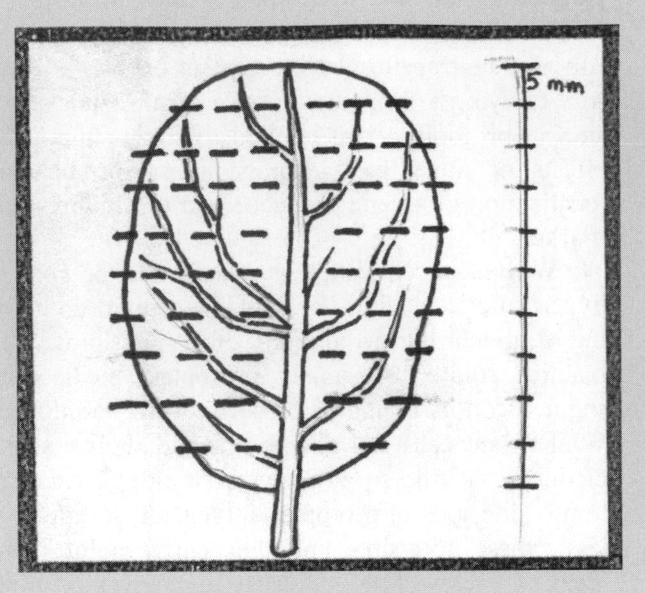

Neutral Red Staining of Vacuoles

1. Transfer 1 mL of protoplast culture into a 1.5 mL microcentrifuge tube, using a wide-bore pipette.
2. Add 1 μL of Neutral Red stain to your sample.
3. Gently invert to mix, wrap the tube in aluminum foil, and incubate for 5 to 10 minutes.
4. Observe cells on the slide without a coverslip, or in a welled-slide with a coverslip. Use a low, dry-objective.

8 OVERVIEW

WORKSHEET

Observations and Assessments

1. Use the space below to sketch your protoplasts, with and without Neutral Red staining. Be sure to include information about the magnification and a scale bar.

8

OVERVIEW

2. Is the staining of vacuoles consistent? Count 25 protoplasts and count the number of darkly-stained, lightly-stained, and unstained vacuoles. Write down these numbers as percentages. Discuss possibilities for the differences in Neutral Red staining.

3. Most of your protoplasts should be perfectly round spheres. Some of them, however, are crenellated or lysed. What conditions would have to exist to cause crenellation? What conditions would have to exist to cause lysis? What can be done to prevent either of these situations from happening to your sample?

4. Your protoplasts were grown in a sterile medium, in the dark under slow, orbital agitation. Explain what would happen if the medium was not sterile. Explain what would happen with the protoplasts were grown in the light. Explain what would happen if the protoplasts were grown under no agitation. Explain what would happen if the protoplasts were grown under higher agitation.

5. What do the chloroplasts look like? From your studies, what do you know about the structure of the typical chloroplast? Describe their structure and draw an example sketch in the space provided below. Feel free to use color in your sketch, if you prefer.

DISCUSSION

Discussion Questions

1. Define apoplast, protoplast, and tonoplast. Discuss the properties of each layer.
2. Imagine an experiment in which you still want to gather seeds from a plant, but you need to create protoplasts to perform an experiment. Discuss possibilities for experimental design that would allow for both.
3. Discuss how staining/testing root tissues might be a possible substitute for protoplast isolation. Would you want young or old roots? How would you grow these plants for ease of experimentation?
4. Protoplasts are kept in a hypertonic buffer, often containing a high amount of mannitol or sorbitol. Why is the buffer hypertonic? Why is the higher osmoticum maintained by one of these sugar alcohols?
5. How long do you think protoplasts can last in sterile media after they are isolated? What causes the collapse of the culture? How could this be prevented? How does this lifespan affect the kinds of experiments you can perform on protoplasts?

OVERVIEW

9 Isolation of Vacuoles

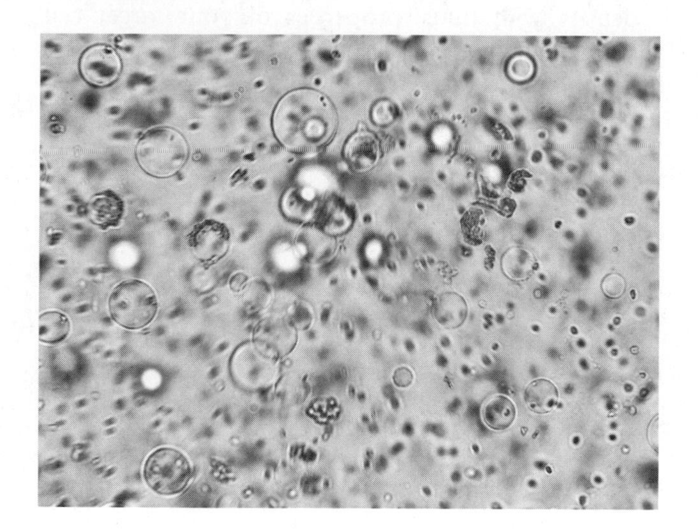

Objectives

- Isolate crude vacuoles from lysis of spinach protoplasts.
- Purify vacuoles using a density-gradient centrifugation.
- Observe crude lysate and purified vacuoles.

DESCRIPTION

Each organelle in the cell has its own unique character. While each organelle is composed of membranes and proteins, the physical and reactive properties of these components differ enough to use them for focused isolation. The density of organelles and vesicles can be used to separate one group of cell components from another, using a density-gradient centrifugation. Heavier components will travel faster through the layers of a sucrose density gradient, for example. Each step of the gradient creates an interface that forms a cushion that less dense structures may settle upon. These components can then be gathered and used for other studies. Today, you will be using a Ficoll-based density gradient centrifugation to separate and purify intact vacuoles from plant protoplasts. First you will lyse the cells, removing the plasma membrane and releasing the cell components. Then you will create a gradient using decreasing concentrations of Ficoll. After centrifugation, your pure, naked vacuoles will be extracted for microscopic observation.

CONCEPTS & VOCABULARY

- Aquaporin
- Density gradient
- Ficoll
- Protoplast
- Tonoplast
- Turgor
- Ultracentrifuge
- Vacuolar H+ATPase
- Vacuole

CURRENT APPLICATIONS

- Density-gradient centrifugation can be used to isolate almost every organelle and vesicle in the cell. Sucrose-density gradients are used to separate by density difference, but the Ficoll-gradient we are using today also separates by solubility in Ficoll. Because the vacuoles are insoluble in the polysaccharide, they rise up to the interface, while other soluble cell components will sink through the gradient. By combining density with other properties of your target cell component, you can isolate at a high level of purity.
- Pure cell compartments and membranes can be used in a variety of studies. The protein components can be analyzed. The activities of enzymes can be tested. The membrane components can be studied. The ability to isolate an organelle is a powerful tool for a cell biologist.

REFERENCES

Karp, G. (2010). Plant cell vacuoles; fractionation of a cell's components by differential centrifugation. *Cell and Molecular Biology: Concepts and Experiments*. (301; 733–734). Hoboken, NJ: John Wiley & Sons, Inc.

Robert S, Zouhar J, Carter C, Raikhel, N. (2007). Isolation of intact vacuoles from Arabidopsis rosette leaf-derived protoplasts. *Nature Protocols*. 2(2):259–262.

BACKGROUND

THE PLANT VACUOLE

A large central vacuole occupies about 30–90% of the volume of most plant cells. Like most organelles, the vacuole is important in maintaining homeostasis. The vacuole is surrounded by a semi-permeable membrane called a tonoplast. Embedded in the tonoplast membrane are various transport proteins that allow for movement of materials into the vacuole or out to the cytoplasm. The aqueous interior of a vacuole is maintained at a low pH. The contents of the aqueous interior include proteins involved in digestion and defense, stored ions and molecules, and most importantly – stored water.

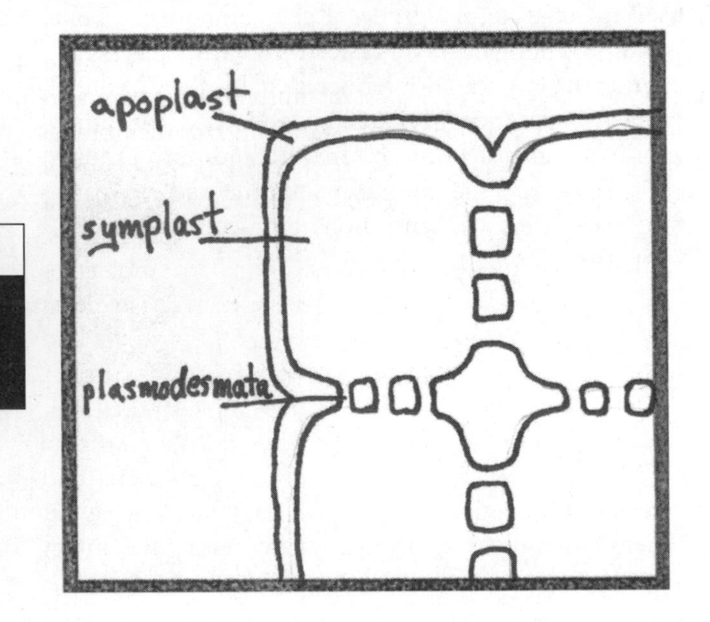

FUNCTIONS

The interior (lumen) of the vacuole contains a number of hydrolases involved in digestion. These hydrolytic enzymes are not activated until they have reached a low pH environment of the vacuole or lumen. Vacuolar proteins are synthesized in the rough endoplasmic reticulum, modified in the Golgi apparatus and sorted for transport at the trans-Golgi network. This distant synthesis and transport of proteins makes the activation by pH difference essential in regulating the activity of these enzymes. In the neutral pH of the ER, Golgi, or cytoplasm, these enzymes would be inert.

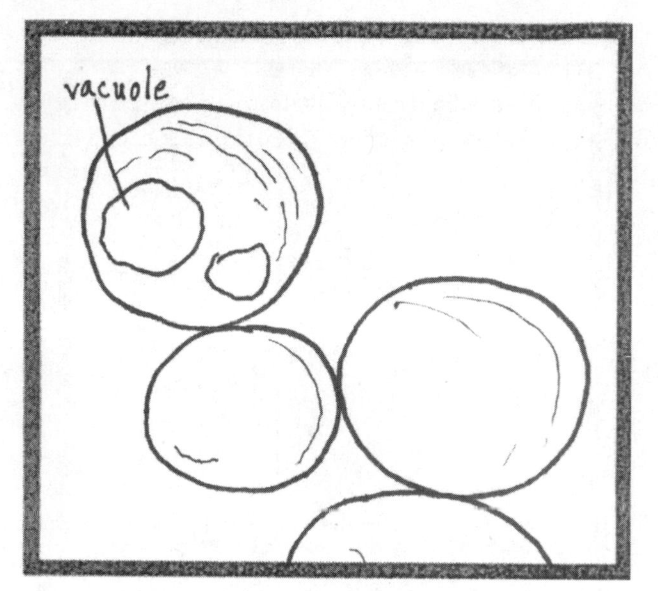

Plant Protoplasts

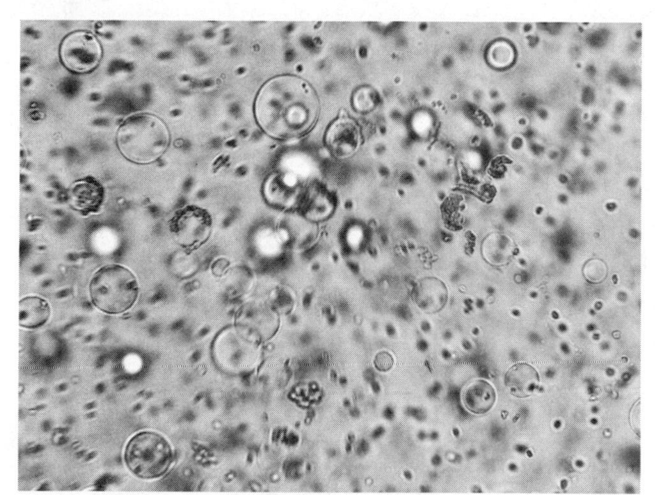
Plant Protoplasts

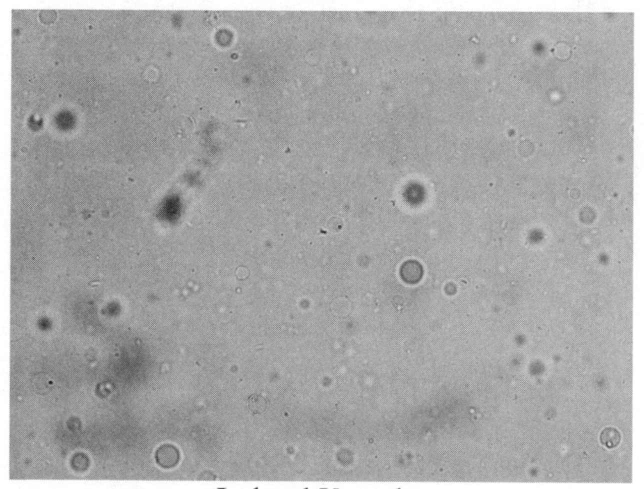
Isolated Vacuoles

The vacuole is also important for storage. As a separate organelle, the vacuole can be used to store toxic compounds important for maintaining defense and immunity. The tonoplast prevents these compounds from being released into the cytoplasm and harming the cell unnecessarily. This separation of compartments also allows the vacuole to maintain the turgor of the cell. Plant cells, due to their stiff cell walls, are able to maintain a higher osmotic pressure compared to animal cells. This state of swollen hypoosmolarity is called turgor. Plants are able to operate at a higher state of efficiency due to the readily available water. In order to maintain this state of turgor, the vacuole acts as a central source for water, ions/solutes, and any other biomolecules the cells need to store for later use. In most adult plant cells, the large central vacuole serves both the roles of digestion and storage simultaneously. However, in the developing cells of the embryo it has been shown that two populations of vacuoles can exist. One population of vacuoles may contain hydrolases (called lytic vacuoles) and the other population may store important nutrients (called storage vacuoles). As the embryo develops, these two populations merge together to provide digested nutrients for the growing plants, and eventually become the large central vacuole.

STRUCTURE

The membrane surrounding the vacuole (called the tonoplast) is composed of a semi-permeable phospholipid bi-layer that is embedded with a variety of transport proteins. The most important of these proteins is the vacuolar H+ATPase, or an ATP-driven proton pump. Using ATP, this pump energetically drives protons (H+) into the vacuolar lumen in order to maintain a low pH. Another important transport protein is aquaporin. Aquaporins transport water in a single direction through a membrane. There are vacuolar aquaporins that pump water into the membrane, and different aquaporins that transport water out of the membrane. Also embedded into the tonoplast are a variety of ion transporters, ion and solute transporters, as well as permeases that allow for traffic of molecules into and out of the vacuoles. These membrane proteins are quickly replaced to allow for efficient turnover and maintenance of the contents of the vacuole.

Today, you will be isolating whole, intact vacuoles. Using detergents, heat, and gentle agitations, you will strip away the plasmid membrane and release the intact vacuoles. You will then use a density gradient centrifugation to purify the intact vacuoles.

PROCEDURES

Protoplast Isolation

1. You will receive a protoplast digestion from your instructor.
2. Sieve digest through 70 nm nylon mesh then pellet cells at 500 × rpm (5′ at 10°C) in a round-bottom tube. Always transfer protoplasts with a wide-bore pipet.
3. Wash 1× with Mannitol/MES buffer.
4. Add Mannitol/MES buffer, and incubate on ice for 30′.
5. Centrifuge to pellet cells, remove all buffer.

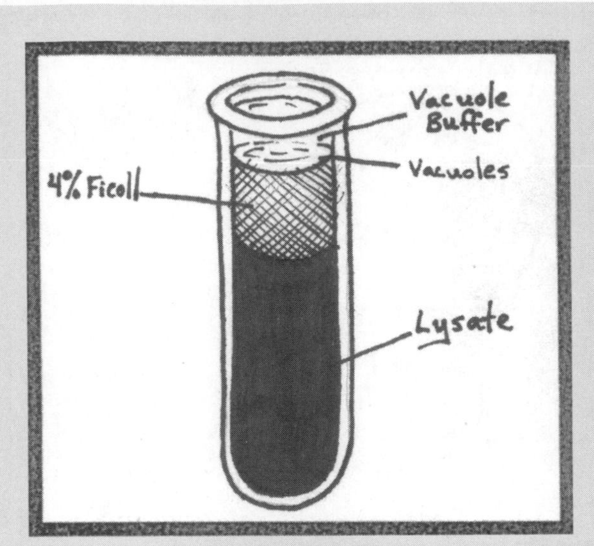

Density Gradient

Vacuole Isolation by Density-Gradient Centrifugation

1. Add 7 mL of pre-warmed Lysis Buffer to protoplasts, mix vigorously by roughly inverting the tube 5 times, and immediately put into ice.
2. Take a small sample of your lysate, and observe it, using the light microscope. Sketch the lysate for question 1 on your worksheet, focusing on the naked vacuoles.
3. Carefully transfer the lysate into a clear Beckman ultracentrifuge tube.
4. Carefully overlay with 3 mL of 4% Ficoll Solution (RT).
5. Carefully overlay with 1 mL of Vacuole Buffer (ice cold).
6. Centrifuge at 50,000 ×g for 50′ at 10°C with slow acceleration/deceleration.
7. Vacuoles can be found in white interface between the Vacuole Buffer and 4% Ficoll Solution. Collect six 200 μL fractions from the top of the gradient, and label them 1 through 6. Your vacuoles should be in fractions 4, 5, and 6.
8. Observe your vacuoles. Sketch the isolate for question 1 on your worksheet, focusing on the naked vacuoles.
9. Continue to observe your vacuoles for 15 minutes. Record your observations in question 2 on your worksheet.
10. Stain your vacuoles with Neutral Red, and record your observations in question 3 on your worksheet.

Ultracentrifuge

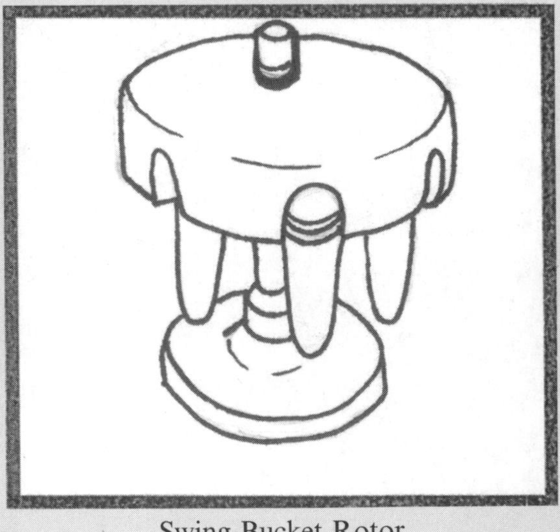
Swing-Bucket Rotor

WORKSHEET

1. Draw examples of the crude lysate and purified vacuole samples that you observe. Describe any visible structures, and draw an example sketch in the space provided below. Remember to label all microscope images with the following: organism and magnification; and add a scale bar at the bottom right of each sketch.

2. Observe your vacuoles for 15 minutes. How does their appearance change? How would you explain this change in morphology?

____

____

____

____

____

____

____

3. Describe your Neutral Red-stained vacuoles. Was the staining uniform? Explain why or why not.

____

____

____

____

____

____

____

4. Explain how the factors of density and solubility in Ficoll were used to isolate vacuoles from the other cell components.

5. You are studying a specific vesicle found in embryonic carrot cells. You know that these vesicles are as dense as vacuoles and insoluble in 2% Ficoll. Design an isolation protocol using the methods that you employed today. Use the space below to explain your protocol, and sketch your gradient setup in the white space.

DISCUSSION

Discussion Questions

1. Discuss the possible tests/assays that you could perform on your purified vacuoles. Some examples include protein studies, membrane composition, analysis of vacuolar contents, and the physical properties of the vacuole.
2. What would you expect to find in the 4% Ficoll layer? What would you expect to find in the Lysis Buffer layer? What would you expect to find in the pellet?
3. What enzymatic activity would you expect to find in tonoplast membranes? In the vacuolar contents? How would you separate the two?
4. How long would you expect the vacuolar-H+-ATPase to function? What would happen to your Neutral Red staining as these pumps start to lose function?
5. Use the Internet to find another organelle isolation method. How does it differ from the vacuole isolation you performed today? What properties of the organelle are being targeted in the method that you found?

Staining of Autophagosomes and Lysosomes

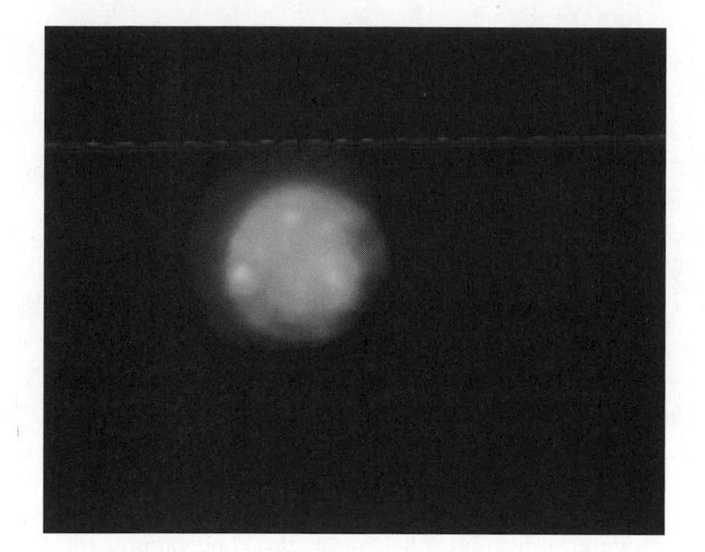

Objectives

- Use monodansylcadaverine to stain autophagosomes.
- Use a lysosome-specific stain to identify lysosomes.
- Compare the fluorescence pattern in a sample stained with both.

DESCRIPTION

Autophagy is a stress-induced mechanism by which the cell can recycle its internal components in a nonspecific manner. Autophagy can be induced by starvation or stress, cell damage, or programmed cell death. The induction of autophagy involves the creation of double membrane-bound structures called autophagosomes. Autophagosomes engulf portions of the cytoplasm including all of their contents to prepare them for digestion. Lysosomes are membrane-bound organelles that contain a variety of digestive enzymes. The lumen of the lysosomes typically has a low pH. In mammalian cells, autophagosomes will bind with lysosomes to form an autolysosome, and the contents of the autophagosome are digested. However, in plant cells after autophagosomes are formed, the lumen of the autophagosome is acidified. Afterward, the autophagosomes are transported to the tonoplast and inserted into the lumen of the central vacuole. Once the outer membrane of the autophagosome has fused with the tonoplast, its inner membrane will be inserted into the lumen and digested. It is through this process that cellular components can be recycled in order to alleviate stress in the cell or redistribute nutrients to the rest of the organism. Today you will be staining autophagosomes and lysosomes and observing them in living cells.

CONCEPTS & VOCABULARY

- Autophagosome
- Autolysosome
- Bulk protein turnover
- Lumen
- Lysosome
- Monodansylcadaverine
- Phagophore
- Tonoplast/vacuole

CURRENT APPLICATIONS

- Typically, autophagy is a non-specific pathway, unlike proteosome-directed protein breakdown, autophagy will break down whatever is engulfed by the autophagosomes, including whole organelles.
- Defects in autophagy has shown to be involved in diseases caused by an accumulation of protein (such as Alzheimer's). Autophagy has also been shown to play an important role in the removal of tumor cells.

REFERENCES

Contento, A, Xiong, Y, Bassham, DC. (2005). Visualization of autophagy in Arabidopsis using the fluorescent dye monodansylcadaverine and a GFP-AtATG8e fusion protein. *Plant Journal.* 42(4):598–608.

Karp, G. (2010). Autophagy. *Cell and Molecular Biology: Concepts and Experiments.* (298–299). Hoboken, NJ: John Wiley & Sons, Inc.

BACKGROUND

DIGESTION WITHIN THE CELL

Just as the cell is the primary source of biosynthesis (anabolic metabolism or anabolism) within an organism, the cell is also the primary site of breakdown of biological molecules (catabolic catabolism or catabolism). Both of these processes require specific enzymes to ensure a rapid reaction. Biosynthetic enzymes may be found within specific organelles, but they may be also found within the cytoplasm. The various hydrolyases and other digestive enzymes that are responsible for catabolic breakdown are sequestered into membrane-bound organelles within the cell. In most fungal and animal cells, small vesicles called lysosomes contain these digestive enzymes. The membrane of the lysosome prevents the enzymes from randomly digesting intracellular components. This membrane also allows the interior (lumen) of a lysosome to maintain a lower pH than the exterior cytoplasm. This lower pH activates the digestive hydrolyases in the lysosome lumen. This activation regulates the activity of the enzymes until they are within the lysosome.

In some fungal and almost all plant cells, lysosomes are not the primary site of digestion. These cells favor a large central vacuole for the storage of digestive enzymes. The vacuole is a large membrane-bound organelle that occupies 30% or more of the cell's interior. The membrane that surrounds the vacuole is called a tonoplast. It is semi-permeable with a variety of transport proteins that allows the passage of water, ions, and other molecules. The vacuole may be a site of storage, but it is also the main site of bulk degradation via autophagy. Autophagy is a typically non-specific process by which portions of cytoplasm and their contents are degraded. Up to 95% of bulk protein breakdown occurs in the vacuole, and organelle breakdown is thought to occur there as well. This makes the vacuole the primary recycling organelle in these cells.

AUTOPHAGY

Autophagy is the primary recycling pathway of the cell. It is induced by nutrient starvation. This may be all nutrients or only one nutrient (for example, nitrogen starvation). Autophagy can also be induced during immune responses or responses to other environmental stresses (for example, oxidative, salt, and drought stresses). In animal and fungal cells, autophagy begins with the formation of the autophagosome. When autophagy is induced, many cup-like membrane structures called phagophores are formed. The phagophore will engulf a portion of the cytoplasm and its contents, and seal itself forming a double membrane-bound autophagosome. Then, the autophagosome will bind with a lysosome, forming an autolysosome, and digest its contents. In plant cells, autophagy is different.

The phagophore will be formed, engulf a portion of cytoplasm, and seal itself to form an autophagosome, but then the interior of the autophagosome will become acidic. The mature autophagosome will travel to the vacuole, where its outer membrane will fuse with the tonoplast. The inner membrane and the contents of the autophagosome will be inserted in the vacuole, where lipases will digest the inner membrane. This will release the contents of the autophagosome, and vacuolar hydrolases will digest the contents.

Today, you will be working with two different fluorescent stains. One stain, monodansylcadaverine (MDC), will specifically label the acidic lumen of mature autophagosomes or autolysosomes. The other stain, Lysotracker, will label small acidic vesicles, including lysosomes and autolysosomes. You will be using the stains to determine and observe the induction of autophagy.

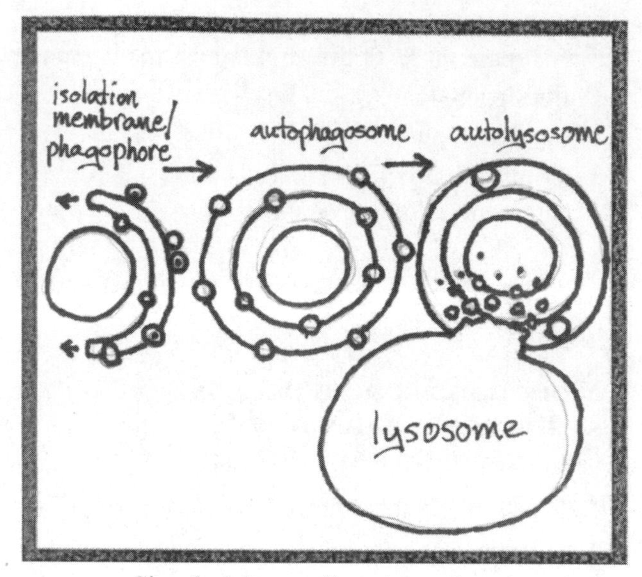

Simple Mammalian Autophagy

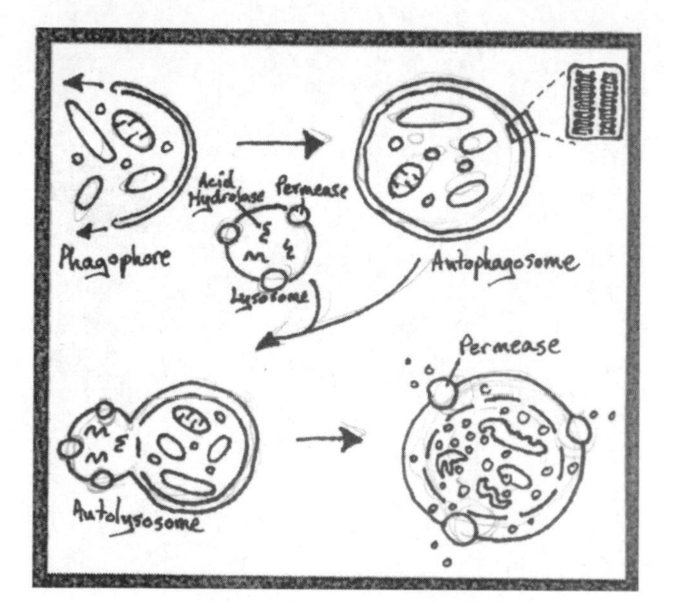

Simple Yeast Autophagy

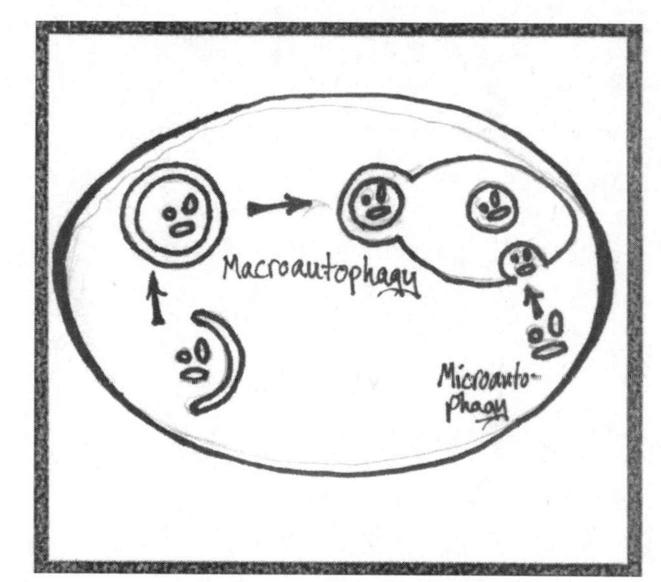

Simple Plant Autophagy

PROCEDURES

Monodansylcadaverine & Lysotracker Staining of Nutrient-Starved Cells

1. Work in the cell culture hood.
2. Obtain two mammalian cell cultures, one control and one that has been nutrient-starved for 2 hours.
3. Remove the medium, and replace it with the MDC-PBS staining solution.
4. Incubate at 37°C for 10 minutes.
5. Remove MDC-PBS staining solution, and replace with 37°C Lysotracker staining solution.

6. Incubate at 37°C for 10 seconds, then remove the medium.
7. Wash 4 times with 37°C 1×PBS.
8. Transfer two drops of cells to a microscope slide, and add a coverslip.
9. Observe each culture with the UV/Light Fluorescence Microscope. Answer questions 1 and 2.

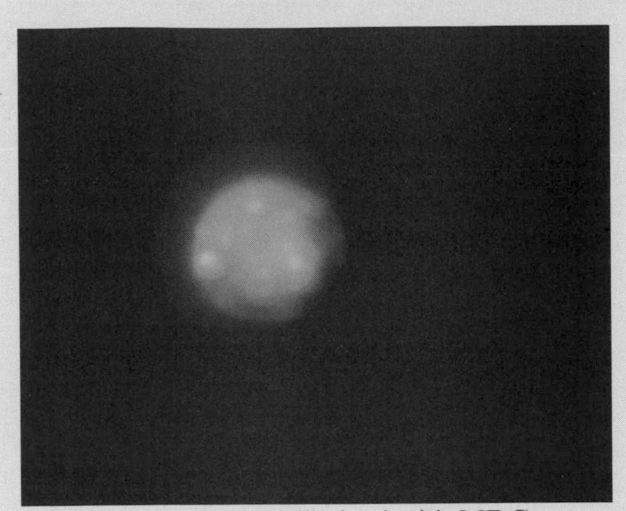

Autophagsomes Stained with MDC

Monodansylcadaverine & Lysotracker Staining of Oxidative-Stressed Cells

1. Work in the cell culture hood.
2. Obtain two mammalian cell cultures, one control and one that has been under oxidative stress for 30 minutes.
3. Remove the medium, and replace it with the MDC-PBS staining solution.
4. Incubate at 37°C for 10 minutes.
5. Remove MDC-PBS staining solution, and replace with 37°C Lysotracker staining solution.
6. Incubate at 37°C for 10 seconds, then remove the medium.
7. Wash 4 times with 37°C 1×PBS.
8. Transfer two drops of cells to a microscope slide, and add a coverslip.
9. Observe each culture with the UV/Light Fluorescence Microscope. Answer question 2 for the oxidative stress samples.

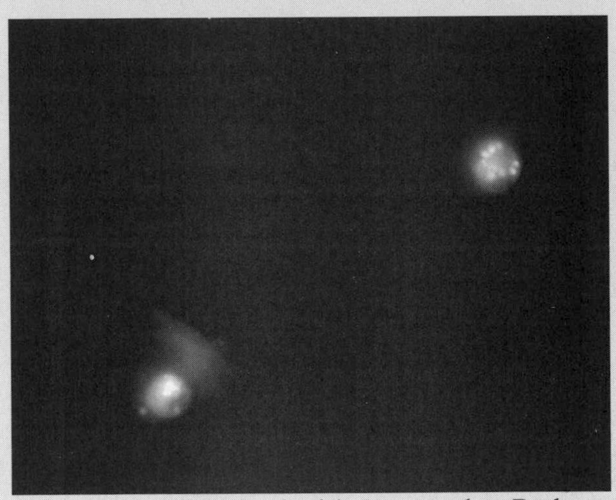

Lysosomes Stained with Lysotracker Red

WORKSHEET

1. What do the control cells look like? What do the starving cells look like? Do you see visible autophagosomes? Do you see visible lysosomes? Describe any visible structures, and draw an example sketch for each sample for the MDC-stained and Lysotracker-stained samples in the space provided below. Remember to label all microscope images with the following: organism, cell type, magnification, and stain; add a scale bar at the bottom right of each sketch.

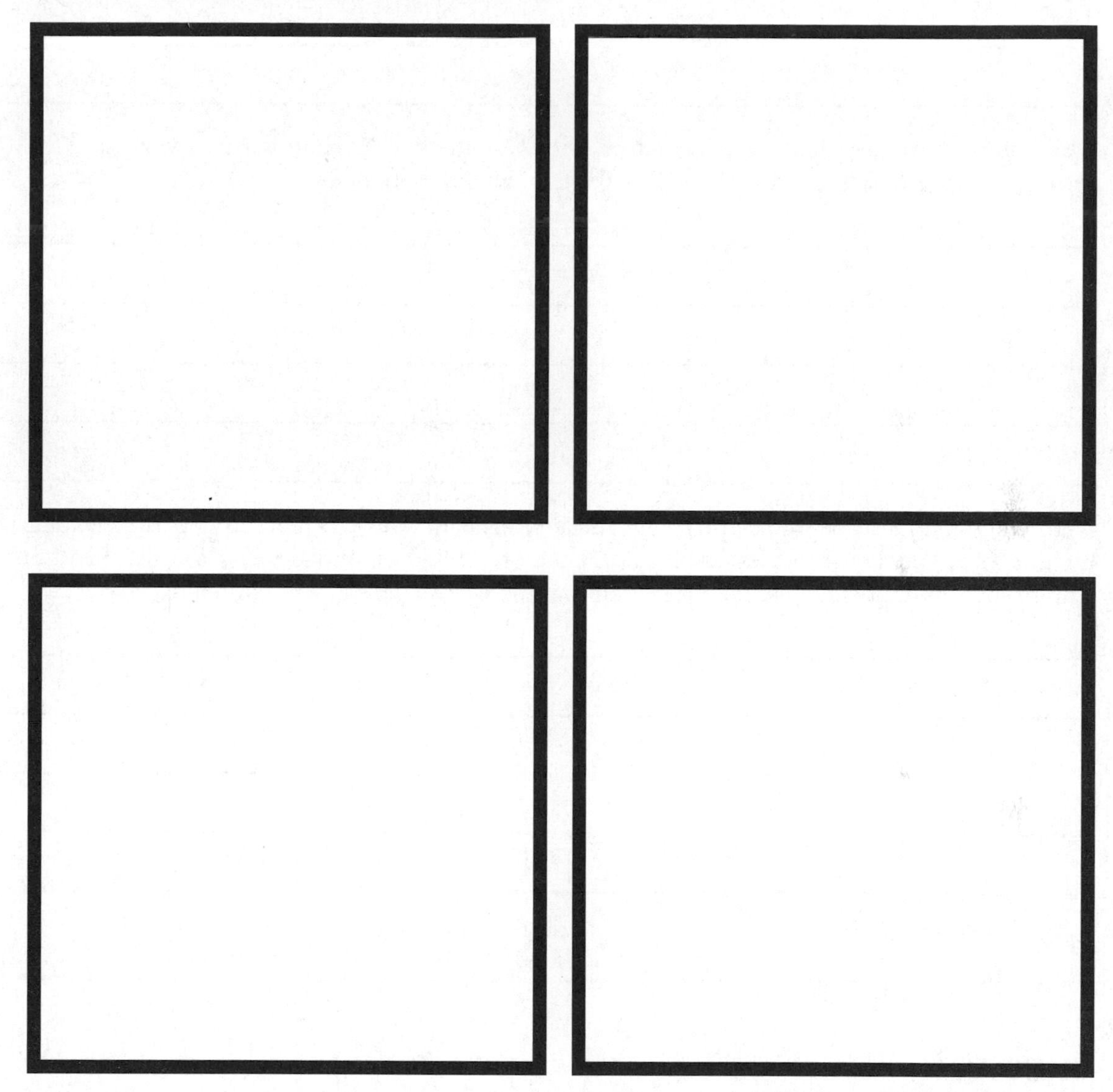

2. Observe 10 cells in each of your samples. Record the data below. Combine your data with the class data and use the class data to answer the two questions that follow.

Sample	Control 1	Starvation	Control 2	Oxidative Stress
# autophagosomes				
Autophagosomes per cell				
# lysosomes				
Lysosomes per cell				
# colocalizations (MDC + Lysotracker				
# colocalizations per cell				

3. Does the MDC-staining colocalize with the Lysotracker staining in any of the vesicles that you observe? Is there more or less colocalization in the starved or oxidative samples, in comparison?

4. Are there any vesicles that are only stained with Lysotracker? Explain why, or why not. Are there any vesicles that are only stained with MDC? Explain why, or why not.

5. Autophagy may also be induced by high salt and drought stress. Design an experiment that would test each of these stresses; then measure for autophagy in cells. You may design an experiment that uses cell culture, but you may also use a living organism and extract cells after treatment.

DISCUSSION

1. Monodansylcadaverine (MDC) and Lysotracker stain the acidic interior cellular compartments. Discuss what other structures in animal or plant cells would be possibly stained by these acidophilic dyes.
2. Lysotracker dyes are dissolved in DMSO. Discuss the reason for using this solvent. What is the function the solvent?
3. Monodansylcadaverine has been shown to have a mild preference for the membranes of autophagosomes, making this stain specific for these structures. The dye itself will only accumulate in the acidic lumen of the autolysosome/mature plant autophagosome. Could this dye label autophagosomes effectively? How would the staining pattern change if you added an ammonium salt to the medium? Explain your answers.
4. Concanamycin A is an inhibitor of the vacuolar H+-ATPase in plant cells. What morphological changes would you expect to see in autophagic plant cells based on what you know about plant autophagy? Explain your answer.
5. Today we induced autophagy by starvation and oxidative stress. When would the cell experience these stresses in a multi-cellular organism?

OVERVIEW 11

Staining of the Cytoskeleton

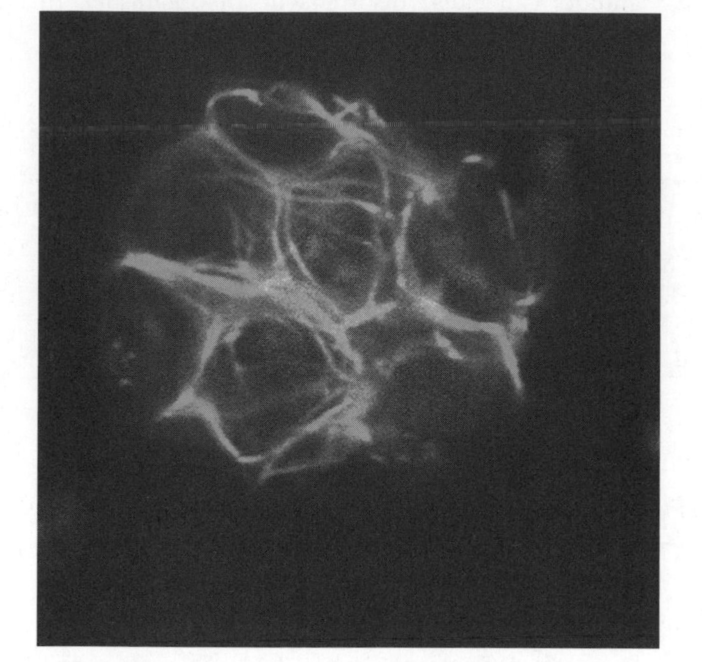

Objectives

- Observe fluorescently labeled cytoskelcton slides.
- Reconstitute the tubulin cytoskeletal elements from a purified sample.
- Stain these elements and observe them.

DESCRIPTION

There are three types of cytoskeleton elements: microtubules, intermediate filaments, and actin filaments. These proteins are important for structural support, intra- and extra-cellular movement, and cellular organization. Microtubules are stiff, hollow cylinders, composed of alpha and beta tubulin. Kinesins and dyneins are two motor proteins associated with microtubules. Actin filaments are flexible, inextensible, helical structures composed of G-actin. Myosin motors bind to actin filaments and move the filaments themselves in an ATP-dependant manner. Intermediate filaments are tough, flexible, extensible, structures of varying composition. While most intermediate filaments are composed of keratins and lamins, there are over 70 proteins that maybe involved in the construction of all the varieties of intermediate filaments. There are a number of ways to label cytoskeleton elements. Today you will be observing labeled microtubules *in vitro* and *in vivo* to study their structure and intracellular location.

CONCEPTS & VOCABULARY

- Actin
- Intermediate filaments
- Keratin
- Lamin
- Microfilaments
- Microtubules
- Phalloidin
- Taxol
- Tubulin

CURRENT APPLICATIONS

- The actin cytoskeleton and associated myosin motors are the key components of muscle tissue. Much of the preliminary research focused on motor proteins, ATP utilization, and calcium signaling was performed on these cytoskeletal proteins.
- Kinesin and dynein motors tow cargo-filled vesicles along the microtubules. These motors are essential for proper intracellular transport. Defects in these motors can cause disorders in all organisms. In humans, cancer, infertility, and even bipolar disorder have been linked to a mutation in these motors.

REFERENCES

Bellocq, C, Andrey-Tonare, I, Paunier Doret, AM, Maeder, B, Paturle, L, Job, Edelstein, D, SJ. (1992) Purification of assembly-competent tubulin from *Saccharomyces cerevisiae*. *Eur. J. Biochem.* 210:343–349.

Harris, JR, Graham, J, Rickwood, D. (2006). Tubulin assembly by taxol (Protocol by Susan L. Bane). *Cell Biology Protocols.* (327). Hoboken, NJ: John Wiley & Sons, Inc.

Karp, G. (2010). The cytoskeleton and cell motility. *Cell and Molecular Biology: Concepts and Experiments*. (318–377). Hoboken, NJ: John Wiley & Sons, Inc.

BACKGROUND

There are three types of cytoskeleton elements: actin filaments, intermediate filaments, and microtubules. These proteins are important for structural support, intra- and extra-cellular movement, and cellular organization. While these structures need to be stable enough to provide support, they are often broken down and reorganized as part as their function. Most of these structures have other proteins associated with them that are important in their maintenance and reorganization, or act as dynamic motors, allowing for movement.

ACTIN FILAMENTS

Actin filaments are flexible, inextensible, helical structures composed of G-actin. Actin filaments are important in maintaining the structure of the cell near the plasma membrane and are important in mitosis. Actin filaments are also important in movements (i.e., muscle movements) and contractility. Actin also has proteins associated with it that are important in reorganization and maintenance, but they burn ATP. Myosin motors bind to actin filaments and move the filaments themselves in an ATP-dependant manner. For example, in a single, muscle fiber, dozens of actin filaments are moved stepwise by hundreds of associated myosin motors.

INTERMEDIATE FILAMENTS

Intermediate filaments are tough, flexible, extensible structures of varying composition. While most intermediate filaments are composed of keratins and lamins, there are over 70 proteins that may be involved in the construction of all the varieties of intermediate filaments. Intermediate filaments are important in structure support, often ending as anchors for other cytoskeleton elements. Like actin filaments and microtubules, intermediate filaments are a dynamic structure, but there are no known motor proteins associated with them.

MICROTUBULES

Microtubules are stiff, hollow cylinders composed of alpha and beta tubulin. The microtubule cytoskeleton is important for structural support in the interior of the cell, intracellular transport, cellular organization, mitotic division, and the primary structure of cilia and flagella. Microtubules have two poles. For labeling purposes one end is called the + end and the opposite end is called the – end. There are a number of proteins that bind to microtubules. Some of these microtubule-binding proteins are important for the

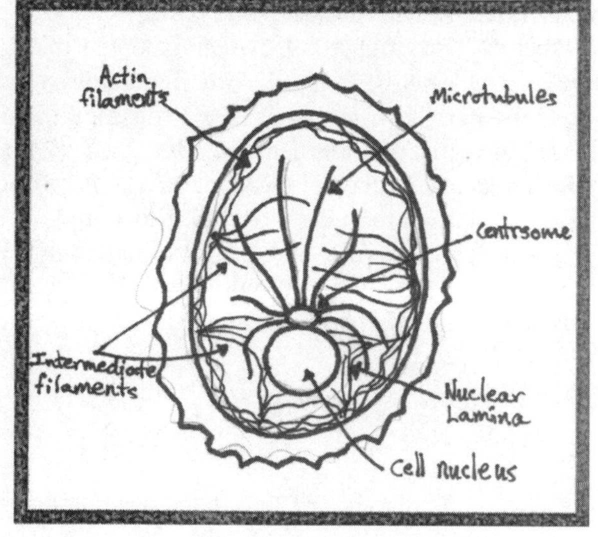

The Cytoskeleton

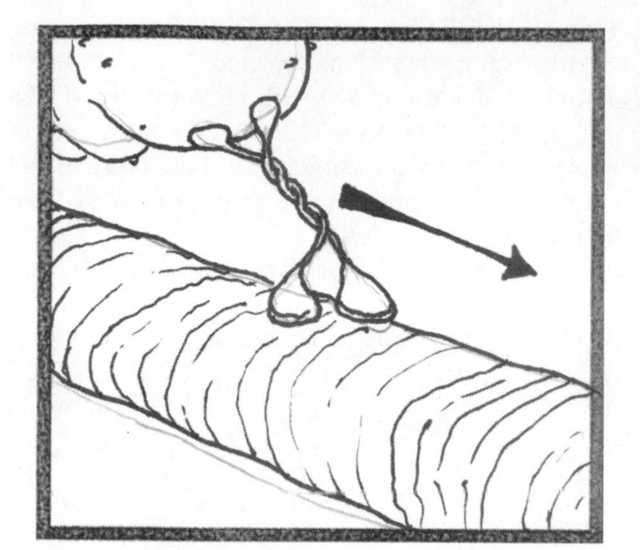

A Kinesin Motor on a Microtubule

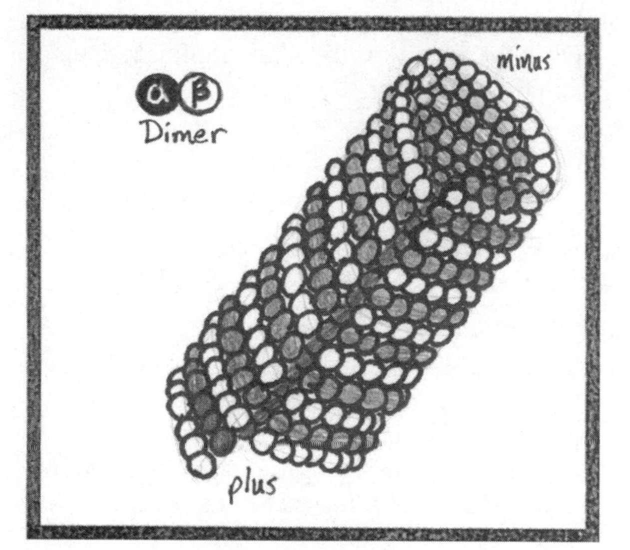

A Microtubule

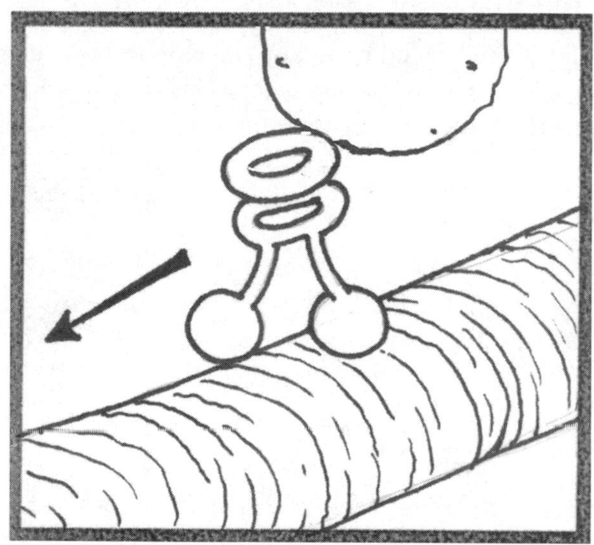

A Dynein Motor on a Microtubule

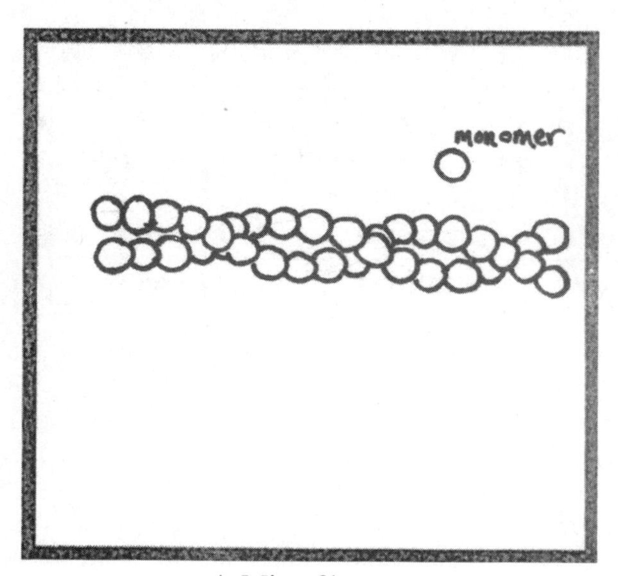

A Microfilament

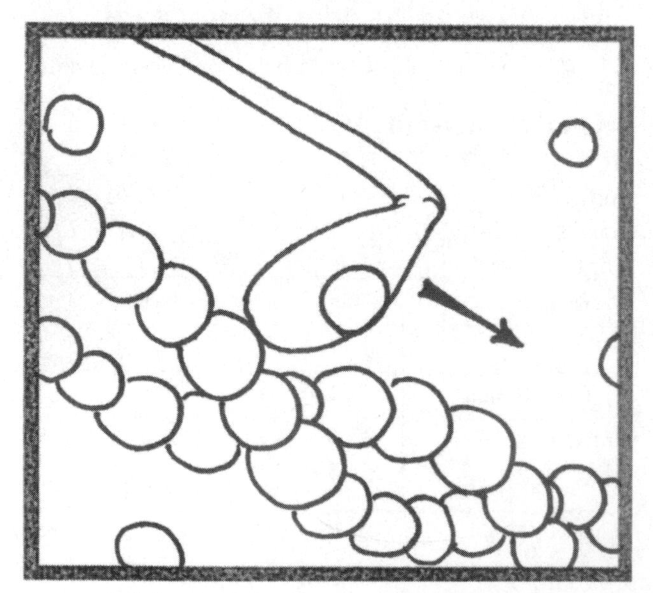

A Myosin Motor on a Microfilament

reorganization and maintenance of the structure, which requires GTP. Kinesins and dyneins are two motor proteins associated with microtubules. These proteins burn ATP to haul cargo along the microtubules. Kinesins haul cargo toward the positive end of a microtubule, and dyneins haul cargo away from the positive end.

There are a number of ways to label cytoskeleton elements. In some cases, the protein component of the element itself can be labeled with a fluorescent tag. The associated proteins and motors may also be labeled. The drug taxol, which binds to microtubules, can also be labeled with fluorescent tags. The same can be done using the drug phalloidin, which binds to actin filaments. Today, you will be observing labeled microtubules *in vitro* and *in vivo* to study their structure and intracellular location. You will also be observing the *in vitro* assembly of microtubules from purified tubulin using fluorescence microscopy and spectroscopic analysis.

PROCEDURES

Observation of Prepared Fluorescent-Labeled Cytoskeleton

1. Obtain a differentially labeled cytoskeleton slide from your instructor. This slide will have fluorescently labeled actin, tubulin or both. Be sure that you have at least one slide for each cytoskeletal element. Observe the slide using a UV-fluorescence microscope and dry/oil objectives (check with your instructor for which filter cubes to use).
2. Observe the actin cytoskeleton. Sketch your observations for question 1.
3. Observe the tubulin cytoskeleton. Sketch your observations for question 1.
4. If available, take a digital microphotograph of each cytoskeletal element and create a merged image (see Appendix A).

Microtubule Assembly by Taxol

1. Obtain the following tubes from your instructor: Buffer A, Buffer G, Taxol A, Taxol F, Tubulin, DMSO.
2. Label four 1.5-mL tubes: Slide, Experimental, No GTP, No Taxol.
3. Set up each reaction using the table below.
4. Keep the reaction tubes at 37°C at all times. Wrap the Slide tube in aluminum foil to keep it safe from the light.

Observation of Microtubule Assembly

1. For the Slide sample, place 30 μL of sample on a slide (using a wide-bore pipette) and cover with a small coverslip. Observe the progress of the microtubule assembly using the high oil-objective on a UV-fluorescence microscope (check with your instructor for which filter cubes to use). Observe at 0, 15, 30, 45, and 60 minutes. Prepare a fresh slide at each time-point. Sketch your observations for question 2.
2. For the Experimental, No GTP and No Taxol samples, measure the A_{350} at 0, 15, 30, 45 and 60 minutes. Transfer the samples to a cuvette using a wide-bore pipette. Keep the cuvette at 37°C at all times.

Sample	Buffer	Tubulin	Taxol	DMSO	dd-water
Slide	50 μL Buffer A	100 μL	0.5 μL Taxol F	5.0 μL	44.5 μL
Experimental	50 μL Buffer A	100 μL	0.5 μL Taxol A	5.0 μL	44.5 μL
No GTP	50 μL Buffer G	100 μL	0.5 μL Taxol A	5.0 μL	44.5 μL
No Taxol	50 μL Buffer A	100 μL	————	5.0 μL	44.5 μL

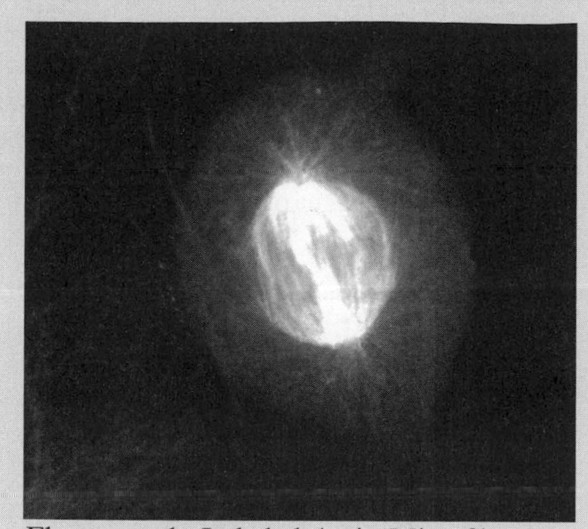

Fluorescently Labeled Actin Microfilaments

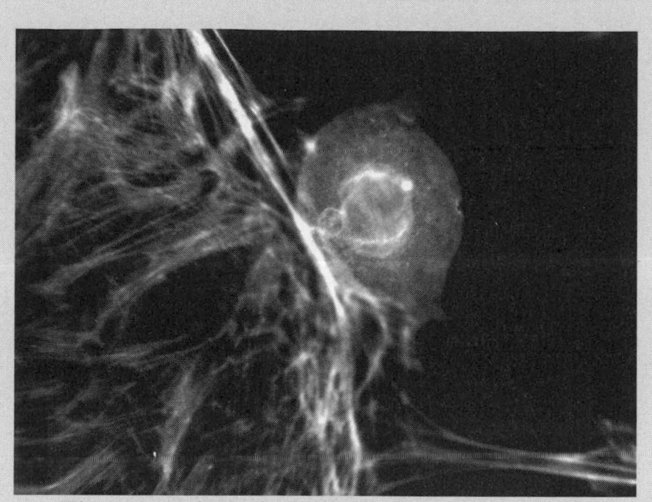

Fluorescently Labeled Tubulin Microtubules

WORKSHEET

1. Observe your pre-stained cytoskeleton slides using epifluorescence. How does the actin look? How does the tubulin look? Draw an example sketch in the space provided below. Using two colors, sketch a drawing with both cytoskeletal elements in the lowest box. Remember to label all microscope images with the following: organism, cell type, magnification, and stain; add a scale bar at the bottom right of each sketch.

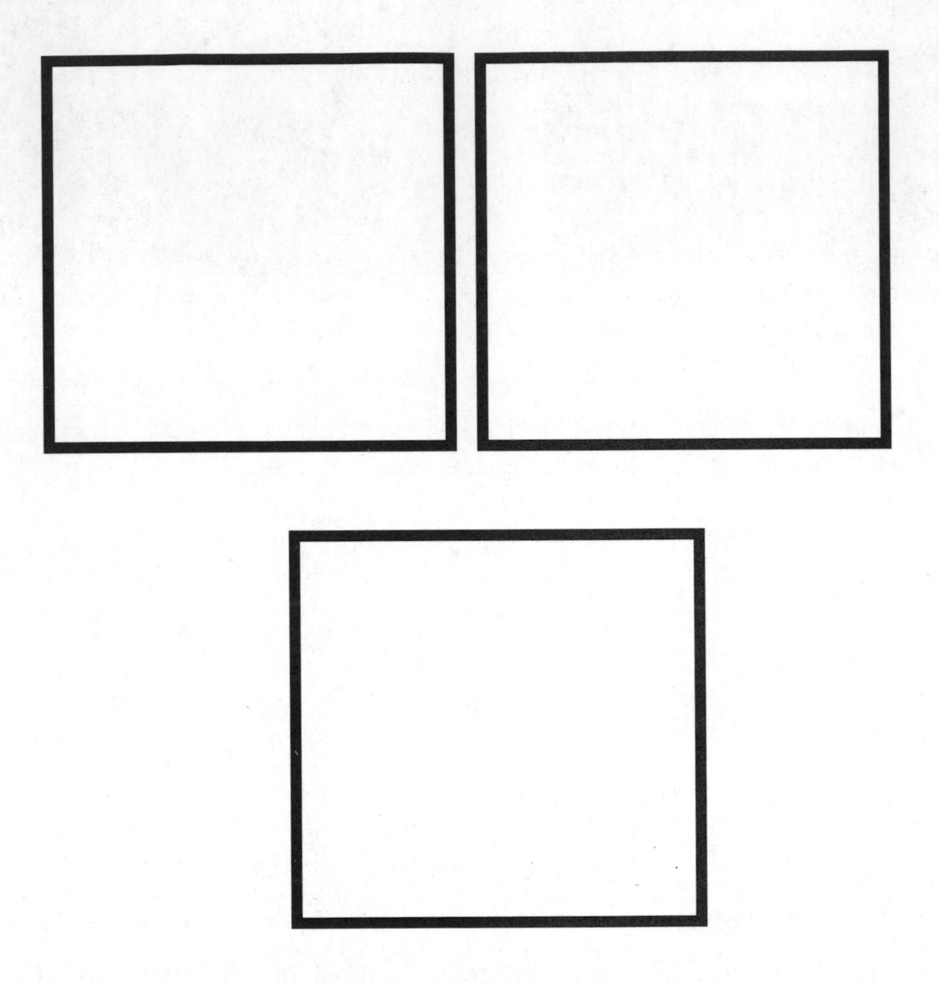

2. Observe your microtubule assembly slide, and sketch the structures that you see at 15, 30, 45, and 60 minutes.

3. Record your A_{350} measurements for your tubulin assembly reactions in the table below.

Sample	0-minutes	15-minutes	30-minutes	45-minutes	60-minutes
Experimental					
No GTP					
No taxol					

4. Your reaction mixture contained taxol and GTP. What is taxol, and how does it promote microtubule assembly? What role did GTP play in the assembly? Why didn't you use ATP for energy? What results would you expect if the reaction mixture was missing taxol? What results would you expect if the reaction mixture was missing GTP? Do your hypotheses match the Absorbance data?

5. You used purified tubulin for your experiment, instead of fresh tubulin extracted from a tissue sample. What results would you expect if you added a crude tubulin sample extracted from a tissue sample to your reaction mixture? Explain your answers.

DISCUSSION

1. What role(s) does the actin cytoskeleton play in maintaining homeostasis in the cell? How are myosins involved?
2. What role(s) does the tubulin cytoskeleton play in maintaining homeostasis in the cell? How are kinesins and dyneins involved?
3. What is meant by a dynamic cytoskeleton? Why is a dynamic actin or tubulin cytoskeleton necessary from proper function of the structure within the cell?
4. Today you used taxol to induce assembly in an *in vitro* sample. Discuss the source of taxol and how it may induce microtubule assembly.
5. You set up two samples today to monitor microtubule assembly. Describe and discuss both methods. Focus on the reaction components and the method of detection.

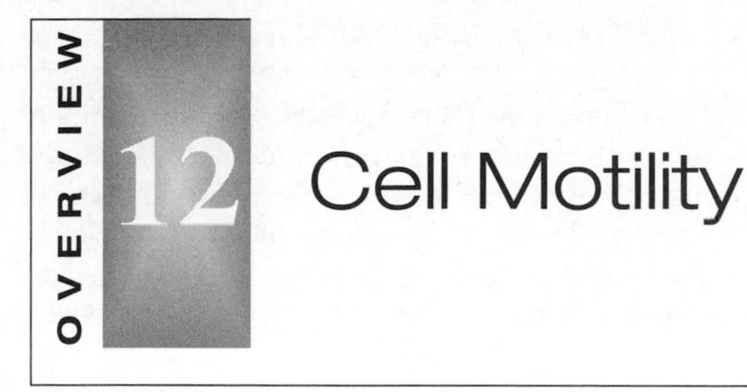

Cell Motility

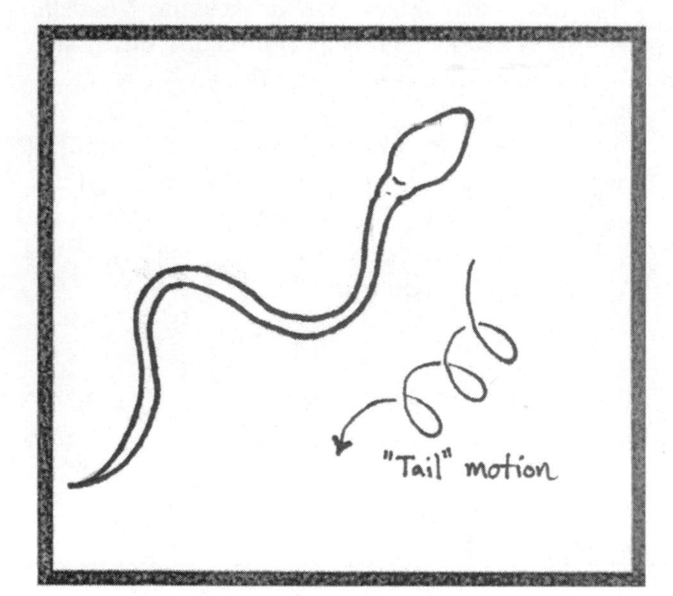

Objectives

- Observe the flagellar motion of Chlamydomonas.
- Observe the ciliary motion of Paramecium.
- Observe the effects of tubulin-motor inhibitors on both of these types of motility.

DESCRIPTION

All cells have some form of motility. This may be intracellular movement along cytoskeletal elements, or it may be extracellular movement that creates movement within a tissue or motility of an entire cell. Cilia are tiny hairs embedded into the membrane into a region known as a basal body. They are typically short and produce motion by acting as whip-like oars, either pushing the cell or moving the fluid adjacent to the membrane. Flagella are a similar structure also connected to the cell by the basal body. However flagella are longer and produce motion in an asymmetric wave form, designed to push the cell through the fluid. Both cilia and flagella are composed of tubulin microtubules and a $9 + 2$ array. The movement of the flagella or cilia is driven by a dynein motor, and transport of materials within the flagella or cilia is controlled by kinesin motors. There are several toxins found in nature that stop cytoskeletal movement. Today, we will use these toxins to directly affect cell motility at the molecular level.

CONCEPTS & VOCABULARY

- AMP-PNP
- Basal body
- BDM
- Cilia
- Dynein
- Flagella
- Kinesin
- Nexin bridge
- 9 + 2 array
- Orthovanadate

CURRENT APPLICATIONS

- Drugs that affect the cytoskeleton are common cancer therapies. Taxol, or paclitaxel, is an extract from the yew tree that causes a permanent stabilization of microtubules. The tubulin cytoskeleton cannot reorganize in the presence of this drug. The drug stops mitosis by stabilizing the mitotic spindles, stopping the rapid division of cells in a tumor.
- Intracellular motility can also be studied using toxins and drugs that affect the cytoskeleton. Vesicular transport and membrane fusion require the tubulin and actin cytoskeletons, respectively. The reorganization of the Golgi apparatus is driven by actin elements. The production of trans-vacuolar strands within the vacuoles of plant cells requires thick actin cables. All of these discoveries were made via experimentation involving drugs that affect the cytoskeleton or motor proteins.

REFERENCE

Karp, G. (2010). The cytoskeleton and cell motility. *Cell and Molecular Biology: Concepts and Experiments.* (318–377). Hoboken, NJ: John Wiley & Sons, Inc.

BACKGROUND

CELLULAR MOVEMENT

Movement is a defining trait for all living things. All cells have some form of motility. This may be intracellular movement along cytoskeletal elements, or it may be extracellular movement that creates movement within a tissue or motility of an entire cell. The most obvious example of motility within a tissue is muscle. Muscle cells are composed of actin filaments, myosin motors, the sarcoplasmic reticulum (which is involved in calcium release and uptake, important for the signaling of muscle movement), and dozens of mitochondria to produce the ATP necessary to drive the process. Another example would be the ciliated pseudo-stratified epithelium, which can contain cilia (tiny hairs composed of tubulin), which push fluids up and out of the esophagus. In many multicellular eukaryotes, the male gamete is a flagellated single cell known as a sperm cell. These flagella are similar in structure to cilia but directly involved in the locomotion of the cell. The final form of movement is amoeboid movement. Amoeboid movement involves the creation of protrusions of cytoplasm called pseudopods. Pseudopods are used to help cells slide into small spaces or pull the cell along the substrate. Examples of amoeboid movements include the infiltration of some types of white blood cells and the family of protists commonly referred to as the amoeba.

AMOEBOID MOVEMENT

In order to create pseudopodia the cell must create cytoplasmic streaming into a small region of the cell. This is achieved by changes in the cytoplasmic density and osmolarity of this region. In some cases, a contractile vacuole is used to create osmotic pressure. Other studies suggest that polymerization and depolymerization of the actin cytoskeleton within the forming pseudo-pod drives the process. Amoeboid movement is a chemotactic process. In most cases, the pseudo-pods are being created to pull the cell in the direction of chemical stimulus. This may be nutrients, a hormone, or some other molecule. This is an energetic process. It does not occur in the absence of ATP. An active and dynamic actin cytoskeleton is also essential for the creation and recovery of the pseudo-pod.

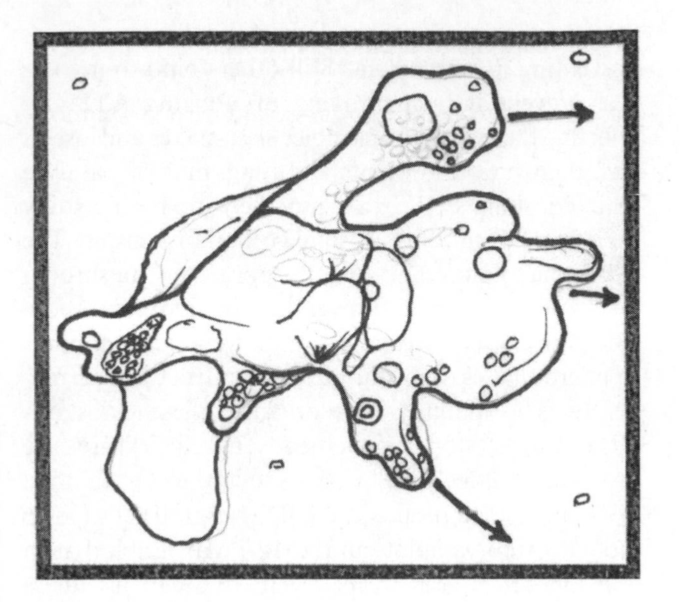

referred to as a doublet. Each doublet contains an A tubule, which is complete and a B tubule, which is incomplete. Each peripheral doublet in the array is bound together by an elastic nexin bridge. All microtubules within the flagella or cilia are aligned in the same direction with the positive ends at the tip and the negative ends near the basal body. The movement of the flagella or cilia is driven by a dynein motor, and transport of materials within the flagella or cilia is controlled by kinesin motors. Both of these microtubule-specific motors require ATP for energy, making this form of movement an active and regulated process.

CILIARY AND FLAGULAR MOTION

Cilia are tiny hairs embedded in the membrane into a region known as a basal body. They are typically short and produce motion by acting as whip-like oars, either pushing the cell or moving the fluid adjacent to the membrane. Flagella are a similar structure also connected to the cell by the basal body. However flagella are longer and produce motion in an asymmetric wave form, designed to push the cell through the fluid.

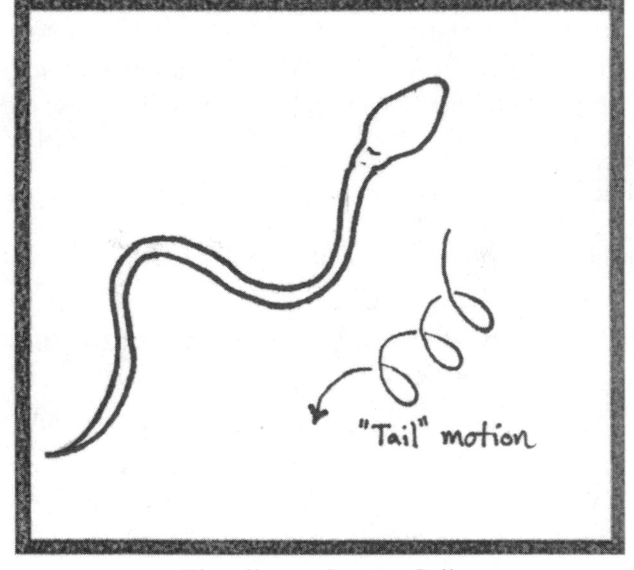

Flagella on Sperm Cell

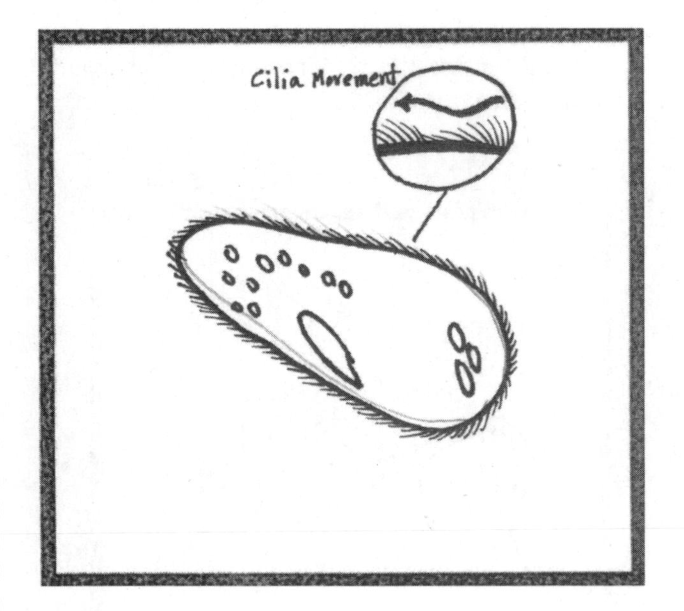

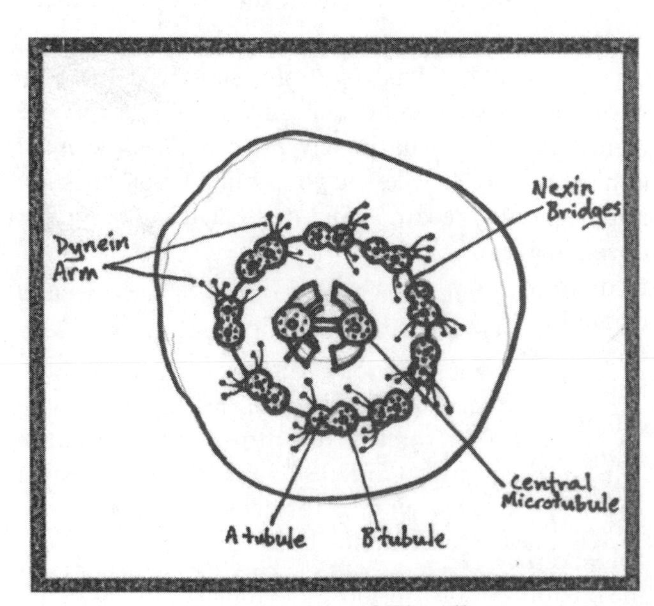

Cross-Section of Flagella

Both cilia and flagella are composed of tubulin microtubules and a $9+2$ array. Two central units are surrounded by a circle of nine peripheral units. Each unit is composed of tubulin structures that are

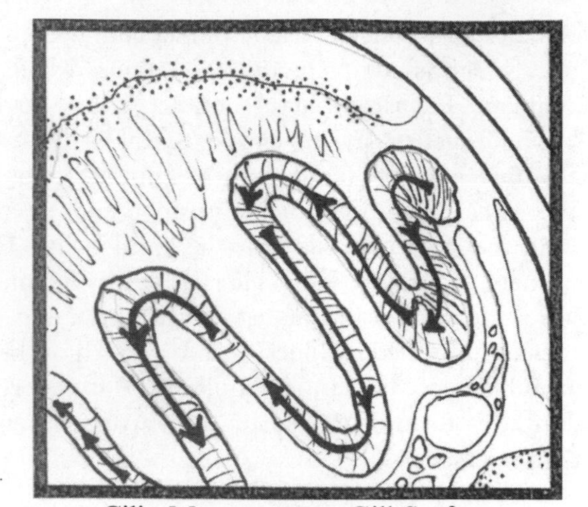

Cilia Movement on Gill Surface

INHIBITION OF CELL MOTILITY

There are several toxins found in nature that stop muscle movement. Some of them do this by removing the calcium signal; others do it by binding to actin – preventing its movement. Still others bind to myosin and prevent it from moving or utilizing ATP. By affecting these ubiquitous molecules, snakes and insects stun their prey and plants and fungi hurt or paralyze their consumers. We can use these toxins to study the nature of motility in single-celled organisms. The toxin phalloidin comes from a poisonous mushroom and binds to actin, preventing its reorganization. The poison, taxol, comes from the yew tree and binds to microtubules, and stabilizes their structures permanently. The inability to reorganize these cytoskeletal elements stops movement entirely. There are drugs that affect the various cytoskeletal motors: 2,3-butanedione monoxime (BDM) paralyzes myosin motors; ortho-vanadate and AMP-PMP disable dynein and kinesin motors, respectively. We can use these toxins and drugs to directly affect cell motility at the molecular level.

PROCEDURES

Observation of Chlamydomonas and Paramecium

1. Obtain a culture of Chlamydomonas and a culture of paramecium.
2. On a clean slide, place a drop of Chlamydomonas culture and add a coverslip.
3. Observe the motion of the organisms. Measure the velocity of the organisms by using the ocular micrometer to measure the distance traveled by 10 organisms for 10 seconds each.
4. Make a new wet mount of Chlamydomonas, and add a drop of Protoslo. Adjust the condenser iris until the flagella are visible. Where are the flagella in reference to the movement of the cell? Sketch your sample for question 1, and record your untreated velocity for question 3.
5. On a clean slide, place a drop of paramecium culture and add a coverslip.
6. Observe the motion of the organisms. Measure the velocity of the organisms by using the ocular micrometer to measure the distance traveled by 10 organisms for 10 seconds each. Sketch your samples for question 2, and record your untreated velocity for question 3.
7. Make a new wet mount of paramecium, and add a tiny drop of Congo red-stained yeast cells with a toothpick. How does the presence of the yeast change the movement of the organism? Is there a visible change in behavior? Do you observe any ingestion of yeast cells? What happens to the color of the pH-sensitive Congo red stain inside the paramecium? Sketch your samples for question 2.

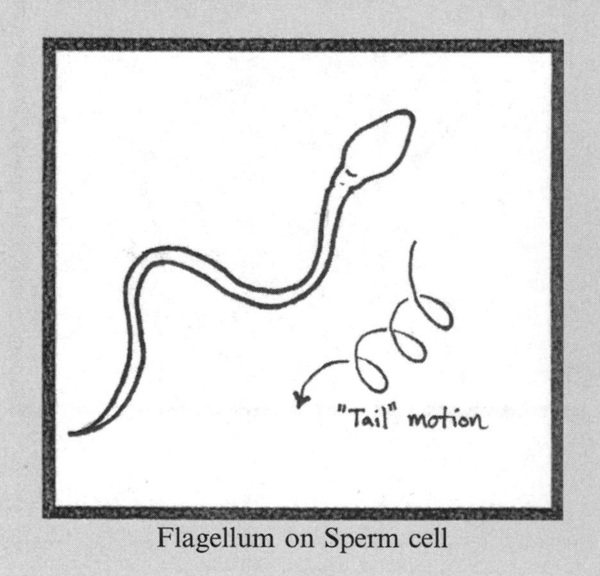

Flagellum on Sperm cell

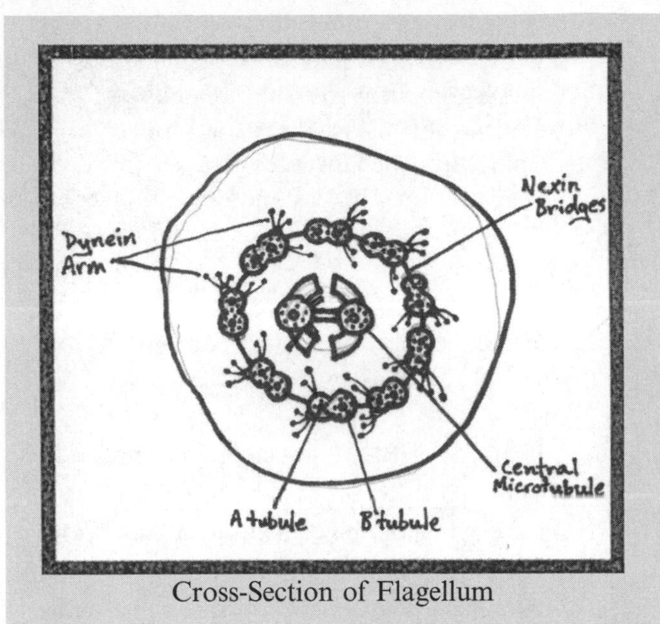

Cross-Section of Flagellum

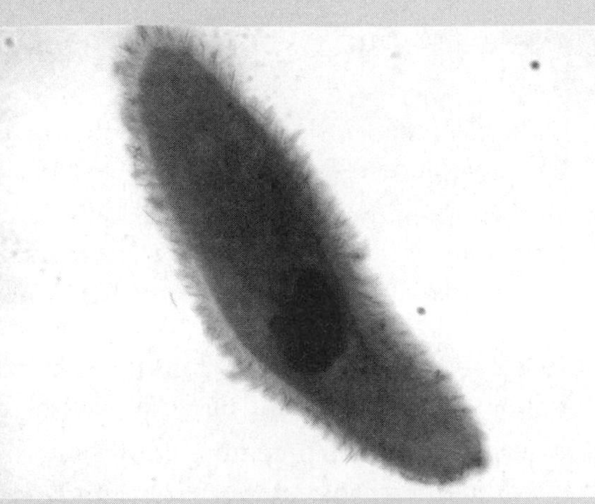
Paramecium

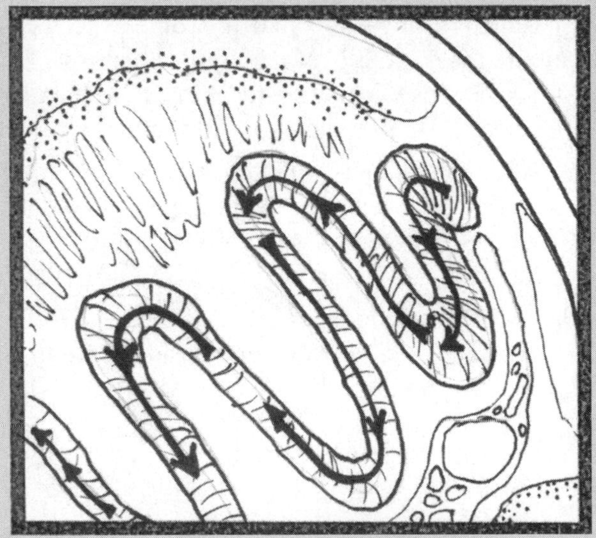
Cilia Movement on Gill Surface

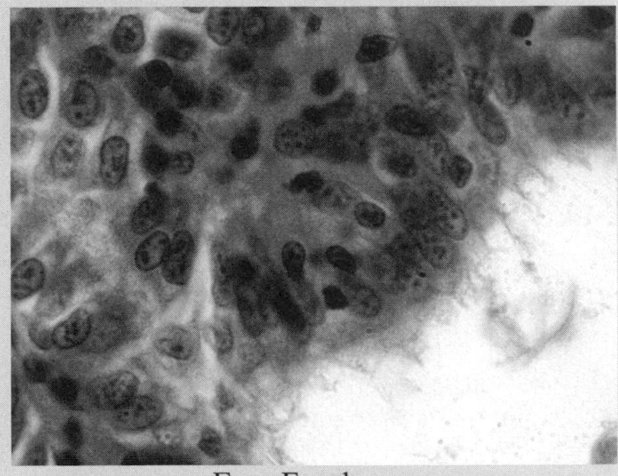
Frog Esophagus

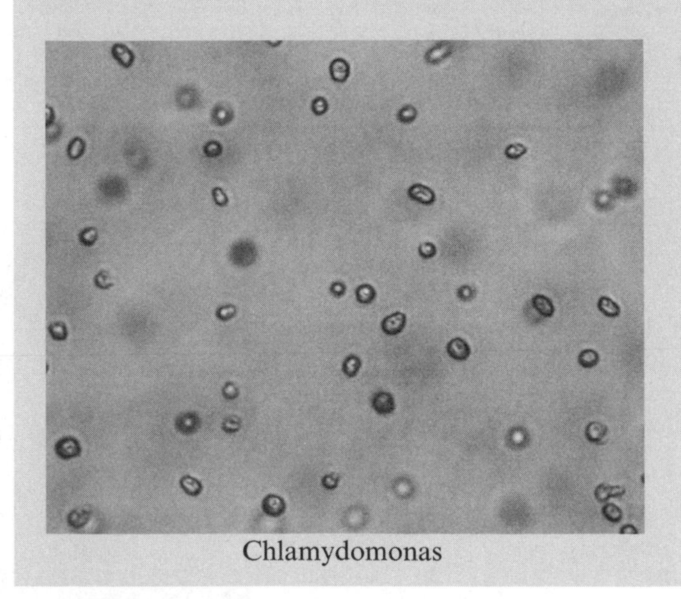
Chlamydomonas

Treatment of Chlamydomonas with Cytoskeletol Motor Inhibitors

1. Separate your Chlamydomonas culture sample into four different tubes in 1 mL aliquots.
2. Label the tubes Control, BDM, AMP-PNP, and Vanadate.
3. On a clean slide, place a drop of the untreated Control Chlamydomonas culture, and add a coverslip.
4. Observe the movement of the cells. After 5 minutes, measure the velocity of the organisms by using the ocular micrometer to measure the distance traveled by 10 organisms for 10 seconds each. Record your untreated velocity at 5 minutes for question 3.
5. After 20 minutes, prepare a new slide, and measure the velocity of the organisms. Record your untreated velocity at 20 minutes for question 3.

6. Add 10 μL of BDM stock to the appropriate tube.
7. On a clean slide, place a drop of the treated Chlamydomonas culture, and add a coverslip.
8. Observe the movement of the cells. After 5 minutes, measure the velocity of the organisms by using the ocular micrometer to measure the distance traveled by 10 organisms for 10 seconds each. Record your BDM-treated velocity at 5 minutes for question 3.
9. After 20 minutes, prepare a new slide, and measure the velocity of the organisms. Record your BDM-treated velocity at 20 minutes for question 3.
10. Repeat procedure for AMP-PNP and Orthovanadate, and record your data for question 3.

Treatment of Paramecium with Cytoskeletol Motor Inhibitors

1. Separate your paramecium culture sample into four different tubes in 1 mL aliquots.
2. Label the tubes Control, BDM, AMP-PNP, and Vanadate.
3. On a clean slide, place a drop of the untreated Control Paramecium culture, and add a coverslip.
4. Observe the movement of the cells. After 5 minutes, measure the velocity of the organisms by using the ocular micrometer to measure the distance traveled by 10 organisms for 10 seconds each. Record your untreated velocity at 5 minutes for question 3.
5. After 20 minutes, prepare a new slide and measure the velocity of the organisms. Record your untreated velocity at 20 minutes for question 3.
6. Add 10 μL of BDM stock to the appropriate tube.
7. On a clean slide, place a drop of the treated paramecium culture, and add a coverslip.
8. Observe the movement of the cells. After 5 minutes, measure the velocity of the organisms by using the ocular micrometer to measure the distance traveled by 10 organisms for 10 seconds each. Record your BDM-treated velocity at 5 minutes for question 3.
9. After 20 minutes, prepare a new slide, and measure the velocity of the organisms. Record your BDM-treated velocity at 20 minutes for question 3.
10. Repeat procedure for AMP-PNP and Orthovanadate, and record your data for question 3.

WORKSHEET

1. What do the Chlamydomonas cells look like? How do they move? Describe any visible structures, and draw an example sketch of the cell and the movement in the space provided below. Feel free to use color in your sketch, if you prefer. Remember to label all microscope images with the following: organism, cell type, magnification, and stain; add a scale bar at the bottom right of each sketch.

2. What do the paramecium cells look like? How do they move? Describe any visible structures, and draw an example sketch of the cell and the movement in the space provided below. Feel free to use color in your sketch, if you prefer. Remember to label all microscope images with the following: organism, cell type, magnification, and stain; add a scale bar at the bottom right of each sketch.

3. Collect your observation data in the table below. Use this data to answer questions 4 and 5.

Sample	Average velocity at 0-minutes	Average velocity at 5-minutes	Average velocity at 20-minutes	Est. time to loss of motility
Chlamydomonas				
Chlamydomonas (BDM)				
Chlamydomonas (AMP-PNP)				
Chlamydomonas (Orthovanadate)				
Paramecium				
Paramecium (BDM)				
Paramecium (AMP-PNP)				
Paramecium (Orthovanadate)				

4. How did BDM affect cell motility? How did AMP-PNP affect cell motility? How did Orthovanadate affect cell motility? Do the effects of these drugs on cell motility match your expectations? Explain your answer.

5. The toxin phalloidin prevents actin reorganization. The drug taxol prevents tubulin reorganization. Using your data from today, describe what effects you would expect from phalloidin treatment of Chlamydomonas or paramecium? What effects would you expect from taxol? Explain your answer.

DISCUSSION

1. Discuss the importance of actin filaments and myosin motors in the movement of cilia and flagella, and amoeboid movement.
2. Discuss the importance of microtubules and kinesin/dynein motors in the movement of cilia and flagella, and amoeboid movement.
3. What are the targets of AMP-PNP, BDM, and Orthovanadate? What are their possible effects on cell motility?
4. What are the targets of phalloidin and taxol? What are their possible effects on cell motility?
5. Taxol slows tumor growth by stabilizing the microtubules of the mitotic spindle and preventing mitosis. Using what you know about the role of the tubulin cytoskeleton, discuss the possible side effects of this treatment in a single cell and within a patient.

Staining of the Nucleus and Nucleic Acids

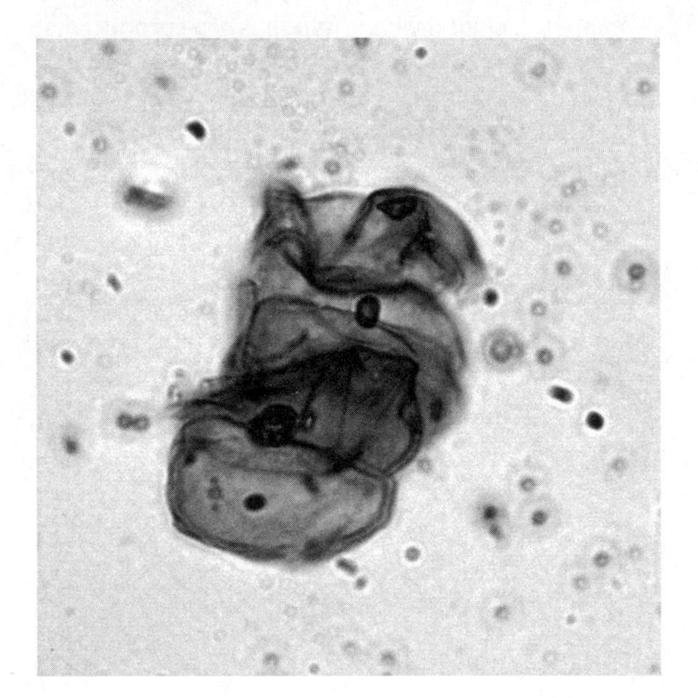

Objectives

- Prepare and observe at least two forms of nuclear staining.
- Identify the key structures of the nucleus.
- Discuss the pros and cons of light versus fluorescent microscopy.

DESCRIPTION

In the majority of cell images, the nucleus is used as a point of reference. Nuclear stains are often used as a counter-stain to help researchers identify individual, intact cells prior to observation of other stains or chromophores. Organelles are described by their proximity to this nearly ubiquitous structure. However, the nucleus has much more to offer than a role as a mere intracellular landmark. The nucleus is the site of genome storage and RNA synthesis. The structure and contents of the nucleus can help a researcher determine if a cell is undergoing DNA synthesis, active mitosis, or a change in gene expression. The condensation of the chromatin, the formation and structure of chromosome, and concentration of DNA or RNA may all be used to determine intimate details of the state of the observed cell. Some staining procedures require only light microscopy, while others require the observation of epifluorescence. We will discuss both of these types of procedures, as well as when they would be best employed.

CONCEPTS & VOCABULARY

- Chromatin
- Chromosome
- DAPI
- Hoechst stain
- Lugol's iodine
- mRNA
- Methylene blue
- Methyl green-pyronin
- Mitosis

CURRENT APPLICATIONS

- In many cases, a chromosomal mutation can be observed by sorting the chromosomes into a display called a karyotype. The most common way to create a karyotype is to take microscope images of all of the chromosomes in a cell. These images may then be sorted by printing them out and cutting and pasting them onto another piece of paper, or by using digital imaging software to perform the same task on a computer. Once the karyotype has been produced, it can be compared to a "normal" karyotype, and differences can be cataloged. Typically, major insertion or deletions can be easily observed.
- The amount of RNA can be quantified in a cell using a variety of staining procedures. Cells undergoing active gene expression will display a higher level of RNA, due to an increase in mRNA, tRNA, and, potentially, rRNA. However, increased expression means increased transcription, not necessarily increased translation. Not all mRNA messages are translated into protein.

REFERENCE

Karp, G. (2010). The cell nucleus and control of gene expression. *Cell and Molecular Biology: Concepts and Experiments.* (475–532). Hoboken, NJ: John Wiley & Sons, Inc.

BACKGROUND

THE ROLE AND MORPHOLOGY OF THE NUCLEUS

The nucleus is the control center of the cell. It contains the majority of the cell's DNA, although mitochondria and chloroplasts do retain some genetic material. This double-stranded DNA (dsDNA) is bound to histones and other DNA-binding proteins to form chromatin, which condenses into chromosomes, the transferable unit of the genome. The nucleus is the site of gene expression, messenger RNA (mRNA) transcript synthesis, and ribosome synthesis and assembly. The organelle is surrounded by the nuclear envelope, which separates the nucleus from the cytoplasm, allowing access to the rest of the cell via nuclear pores and other membrane-bound transporters. Within the nucleus, one region that can typically be identified is the nucleolus, which is the site of ribosomal RNA (rRNA) synthesis and ribosome assembly. The structure of the nucleus is maintained by the nuclear lamina, composed of intermediate filament proteins that primarily bind to the nuclear envelope and maintain chromatin integrity.

The nucleus is ubiquitous in almost all cell types. Some cell types, like erythrocytes (red blood cells), are anucleated if they are mature. During development the embryos of some organisms can form a multinucleated cell called syncytium. In the fruit fly Drosophila, for example, nuclear division occurs rapidly within the fertilized egg cell, without cytokinesis. Eventually, each nucleus is surrounded by a plasma membrane within a portion of the cytoplasm to create many smaller cells, each with a single nucleus. It is the presence of a nucleus that can be used to define a unique cell, as opposed to an extracellular compartment or structure. Intact DNA and chromosomes can be used to identify healthy, living cells, since DNA degradation is a common feature of dying or dead cells. The location of the nucleus can also be used to identify the location of other structures in the cell, like the endoplasmic reticulum, which is always in contact or close proximity with the nucleus.

STAINS USED TO OBSERVE THE NUCLEUS

Today, you will use several stains to highlight the structure and/or contents of the nucleus. There are many stains that specifically bind to DNA, like the fluorescent Hoechst and SYTO stains, which is the easiest way to locate the nucleus. There are several basophilic stains that accumulate in the nucleus due to its higher pH, like hematoxylin, which will stain the nucleus blue. Then there are some stains that just act as contrast within the cell, allowing for easier observation of the nucleus by giving some color to an otherwise clear cell. *Lugol's iodine* will stain starch or glycogen in the cell, and make the nucleus more visible without actually staining the nucleus itself.

Methylene blue binds to nucleic acids and can be used to stain the nucleus with a visible blue color. Methylene blue is non-toxic and can be used safely, although it will stain skin and clothes. DAPI is a fluorescent stain that also binds tightly to nucleic acids. Because of this stronger binding, DAPI is toxic and mutagenic. However, it is a commonly used counterstain for the nucleus in fluorescence microscopy. DAPI must be excited by an ultraviolet light source and has an excitation/emission peak of 358nm/461nm when it binds to dsDNA. DAPI is often used in concert with fluorescent stains or proteins with different excitation/emission spectra. For example, green fluorescent protein (GFP) is a common label for recombinant proteins, but has a unique fluorescent signal when compared with DAPI. These two chromophores can be used together in the same sample for microscopic observation, or for flow cytometry experiments.

Flow cytometry uses fluorescent labels to count the number of cells with fluorescence by volume in a sample. DAPI staining could be used to determine the total number of intact cells, or the total number of living cells, depending upon the experiment. Flow cytometry is a useful technique, since it saves the time and effort of counting the number of stained cells on a slide under the microscope. This technique can also measure the amount of fluorescence, allowing for quantification of the amount of DNA, for example. Flow cytometry can even be used to identify different types of cells. Most hematology analysis, the differentiation and quantification of red and white blood cells, is done by flow cytometry.

PROCEDURES

Staining of the Nucleus and Nucleic Acids

Staining of Fresh Onion Root Tissue

1. Add a drop of sterile water to the center of a clean microscope slide.
2. Using a razor blade carefully cut a fresh root from the basal plate.
3. Transfer the root to your slide, and place it in the water drop. Add another drop of water and cover with a cover slip.
4. Using the eraser-end of a pencil, gently crush the root underneath the cover slip until the cells spread into a single layer.
5. Examine the sample under the microscope with low- and high-power objectives, and make a note of the appearance on your datasheet. Answer the first part of question 1 on your data sheet.
6. Add 1–2 drops of *Lugol's iodine* to one edge of the cover slip. Encourage diffusion of the stain by applying a small piece of paper towel to the opposite end of the cover slip. Take care not to completely dry out your slide!
7. Examine the sample under the microscope with low- and high-power objectives, and make a note of the appearance on your datasheet. Look for the nucleus, and examine its contents. What is stained? Can you see chromatin or chromosomes in any of the cells? Can you recognize any other structures? Answer question 1 on your datasheet.

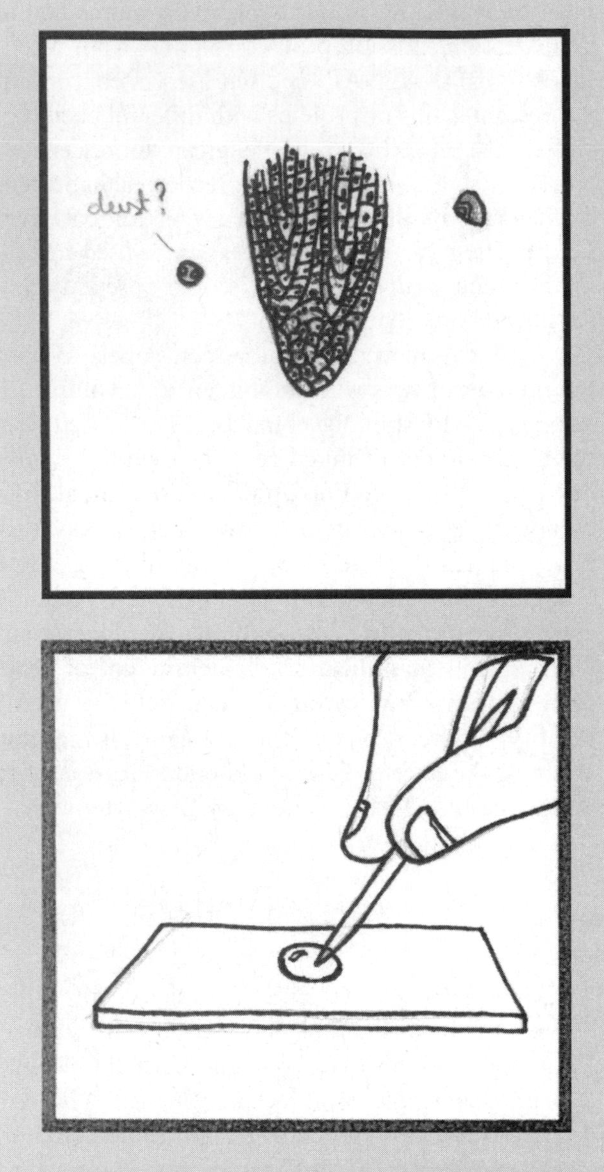

Staining of Mammalian Epidermal Cells with Methylene Blue

1. Put a drop of 0.3% methylene blue onto a microscope slide. **Caution: methylene blue will stain clothing and skin.**
2. Gently scrape the inside of your cheek with the flat side of a toothpick.
3. Stir the end of the toothpick in the stain, and discard the toothpick.
4. Place a cover slip on the slide.
5. Examine the sample under the microscope with the low-power objective, and make a note of the appearance on your datasheet. Cells should appear to be clear or purple blobs.
6. Once you have located a cell, switch to a high-power objective, and examine the nucleus. Answer question 2 on your datasheet.

Staining of Mammalian Epidermal Cells with DAPI

1. Using a sterile toothpick, scrape the inside of your cheeks.
2. Pour 10 mL of 1× phosphate-buffered saline (1X PBS) into your mouth and swish the buffer around for 5 seconds.
3. Expectorate the buffer into a paper cup, and then transfer it back to the 15 mL conical tube that held your 1× PBS. Cap the tube and wipe off the exterior.
4. Pellet the cheek cells in a clinical centrifuge on setting 2 for 10 min.
5. Remove the supernatant, and resuspend the pellet in 5 mL of DAPI staining buffer.
6. Incubate for 15 min at room temperature in the dark.
7. Pellet the stained cheek cells in a clinical centrifuge on setting 2 for 10 min.
8. Remove the supernatant and resuspend the pellet in 5 mL 1× PBS.
9. Add one drop of your cell suspension to a microscope slide.
10. Place a cover slip on the slide.
11. Examine the sample under the light microscope with the low power objective, and make a note of the appearance on your datasheet.

Cells should appear to be clear blobs. Switch over to DAPI-fluorescence, and observe the staining. Make note of the appearance on your datasheet.

Cheek Cell Nucleus

12. Once you have located a stained, fluorescent cell, switch to a high-power objective and examine the nucleus. Answer question 3 on your datasheet.

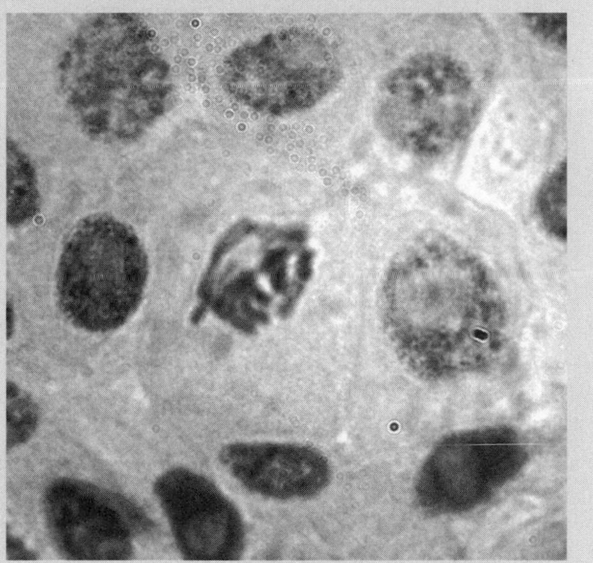

Root Tip Nuclei

WORKSHEET

Observations and Assessments

13

OVERVIEW

1. What do the unstained onion cells look like? Describe any visible structures and draw an example sketch in the space provided below. Feel free to use color in your sketch, if you prefer. What do the stained onion cells look like? Can you see the nucleus now? Can you see the chromosomes? Can you see any cells that are going through mitotic division? What is the iodine staining, and how does this help you to visualize the nucleus? Describe any visible structures and draw an example sketch in the space provided below. Remember to label all microscope images with the following: organism, cell type, magnification, and stain; add a scale bar at the bottom right of each sketch.

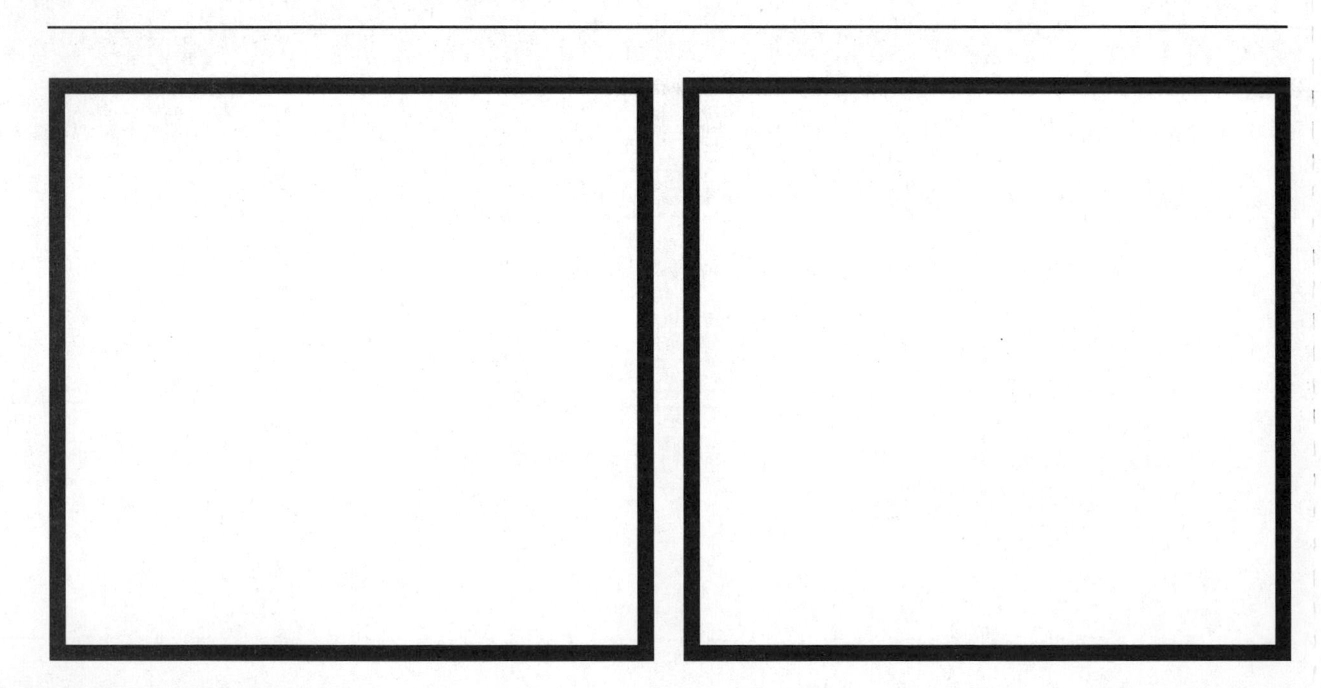

2. What do the unstained cheek cells look like? Describe any visible structures and draw an example sketch in the space provided below. What do the stained cheek cells look like? Can you see the nucleus now? Can you see the chromosomes? Can you see any cells that are going through mitotic division? What is the methylene blue staining, and how does this help you to visualize the nucleus? Describe any visible structures and draw an example sketch in the space provided below.

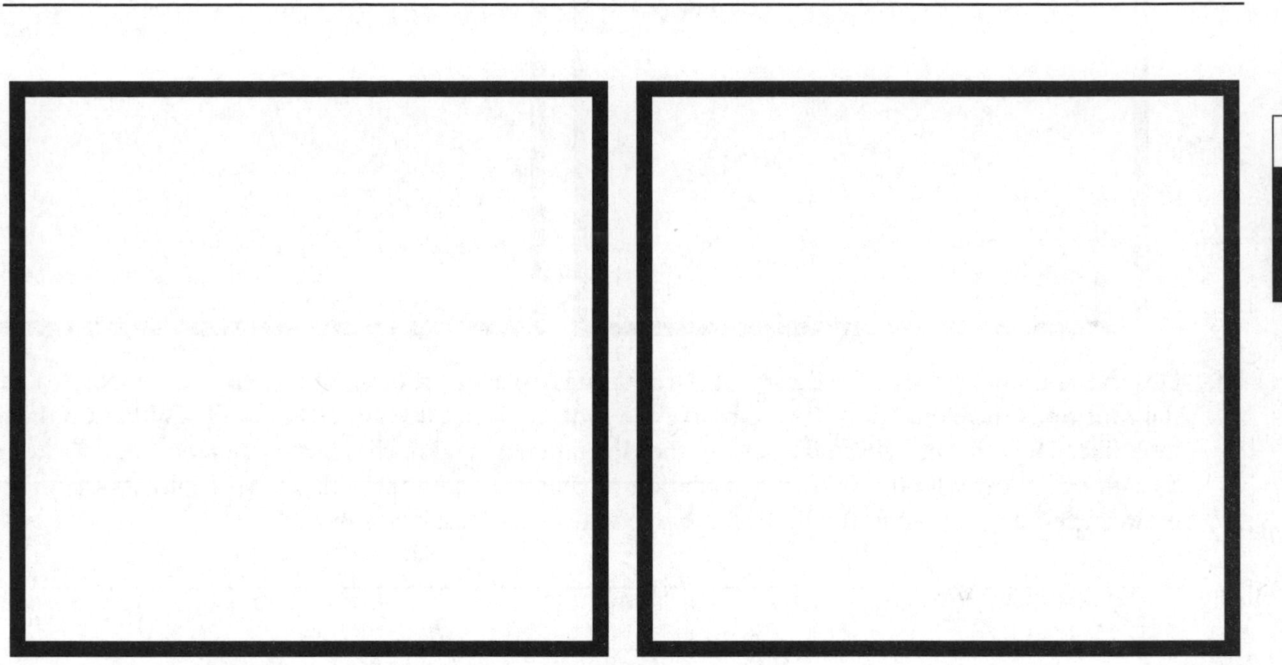

3. What do the DAPI-stained cheek cells look like? Can you see the nucleus now? Can you see the chromosomes? Can you see any cells that are going through mitotic division? What is the DAPI staining, and how does this help you to visualize the nucleus? Describe any visible structures, and draw an example sketch in the space provided below. Draw an image of the nucleus observed under the high-power objective, and label any nuclear structures that you recognize.

4. Observe your onion roots. The region at the bottom is called the root tip or meristematic zone. Just beyond the root tip is the elongation zone. Observe the root tip. What do the cells look like? What do the nuclei look like? How do the cells and nuclei in the elongation zone differ? Which population of cells are going though active cell division? Which population of cells are going through growth? Discuss the difference between cell division and growth, and the role that the nucleus plays in each.

____

____

____

____

____

____

____

5. What are the benefits of fluorescence microscopy? What are the shortcomings? Discuss when you should use light microscopy versus when you should use fluorescence microscopy.

____

____

____

____

____

____

____

DISCUSSION

Discussion Questions

1. Discuss the targets of Lugol's iodine, methylene blue, and DAPI. How does each stain assist in the visualization of the nucleus and its contents?
2. Onion root cells and human epidermal cells were chosen because these tissues contain actively dividing cells. What effect does active cell division have on the structure of the nucleus? What would the nuclei of non-dividing cells look like? Discuss other possible sources actively-dividing tissues in plants and people.
3. Discuss the safety precautions that must be remembered when using methylene blue and DAPI. Should safety be a factor in the choice of a staining procedure?
4. Discuss the differences between methylene blue and DAPI staining. What does each compound stain? What structures are stained better by each stain? When would you employ each stain?
5. DAPI stains double-stranded DNA and single-stranded RNA differently. For DAPI complexed with dsDNA, the emission peak is 461 nm. For DAPI complexed with ssRNA, the emission peak is 500 nm. Using the background information in this chapter, discuss a possible strategy to measure the amount of ssRNA in a cell sample.

Staining of Chromosome and Cell Cycle Analysis

Objectives

- Prepare and observe stained chromosomes in fava bean roots.
- Identify the key structures of the chromosome.
- Identify the types of chromosomes and the stages of mitosis.
- Quantify cells in each stage of the cell cycle within a Feulgen-stained root sample.

14

OVERVIEW

DESCRIPTION

Within the cell, DNA is a master blueprint for the entire organism. The expression of RNA, transcribed from the DNA code, determines the nature of each cell. While the DNA remains the same in every cell, RNA molecules found in the cytosol can vary from cell type to cell type. The ability of genes to be expressed selectively gives an organism almost infinite flexibility and variety when producing cell morphology and physiology. Of the 30,000 genes found in the human genome, only a few thousand are expressed in a cell at any given time. From cell to cell, only a few hundred genes will be expressed in all cells. Control over gene expression is a molecular and a physical process. Certain genes can be selected for transcription by various transcription factors, but the structure of the chromosome also plays a vital role in controlling gene expression. The tightly-wound DNA supermolecules are designed to unwind and reveal specific gene loci, thus allowing for a secondary physical control over what genes are allowed to be expressed in each cell type. When cells divide, chromosomes are duplicated and divided between two daughter cells. This compact packaging also insures that all genetic information is transferred to the new cell with each cell division.

CONCEPTS & VOCABULARY

- Euchromatin
- Heterochromatin
- Metacentric
- Acrocentric
- Telocentric
- Centromere
- Secondary constriction
- Chromatids
- Fast Green FCF
- Feulgen Reaction

CURRENT APPLICATIONS

- Some cells contain more DNA or RNA than others. Cells that are actively dividing will often contain more DNA than quiescent cells. Cells that are actively expressing large amounts of protein, like the cells of the pancreas or liver, will often contain more RNA than cells found in muscles or nervous tissues.
- Histones and DNA-binding proteins allow DNA to condense into chromatin and chromosomes. Without these proteins to modify the structure of DNA, the nuclei of cells would have to be 20 times larger to hold the entire expanded genome. There are several diseases linked to defects in histone function, included some types of Alzheimer's disease and Huntington's disease.

REFERENCE

Karp, G. (2010). The cell cycle and mitosis. *Cell and Molecular Biology: Concepts and Experiments.* (560–589). Hoboken, NJ: John Wiley & Sons, Inc.

BACKGROUND

In order for multi-cellular organisms to grow and develop, the processes of cell division and cell expansion must occur in concert. Cell division involves the tightly regulated synthesis and separation of new genetic material coupled to the division of cytoplasm. In most cases, cell division converts a single cell into two smaller daughter cells, each with a full complement of chromosomes and cytoplasmic contents, contained by their own distinct plasma membrane. As these daughter cells mature, they will increase their size and cytoplasm as they perform their programmed anabolic processes. This cycle of division and expansion is common to most cells in every organism.

The cell cycle is punctuated by several phases. Interphase is the common state of the cell cycle, and contains the Gap and S-phases. Gap phases represent the periods of growth and the metabolism associated with the cell's specific function. Most cell cycles have two Gap phases, separated by an S phase. The S phase is the period of new DNA synthesis, where the entire genome is copied in preparation for mitosis. Mitosis is the period when genetic material condenses into chromosomes and is distributed between the two daughter cells. Mitosis itself is divided into six stages: Prophase, Prometaphase, Metaphase, Anaphase, Telophase, and Cytokinesis. Figure 14.11 in your Karp text has an excellent breakdown of the key steps in mitosis. Today, you will be observing and identifying the phases of the cell cycle and the steps of mitosis by studying the actively dividing cells of a Fava bean root.

CHROMATIN, THE STRUCTURE OF CHROMOSOMES AND CELL DIVISION

Double-stranded DNA is stored in the nucleus of the cell in the form of chromatin. Chromatin is a complex of DNA wrapped around histone proteins and other DNA-binding proteins. This bundling of DNA reduces the space necessary to store the cell's genetic material. Regions of chromatin may be tightly packed into heterochromatin, or loosely packed into euchromatin. The more loosely packed euchromatin is open and available for mRNA transcription and gene expression.

During mitosis, the chromatin condenses even further into chromosomes. Chromosomes are tightly

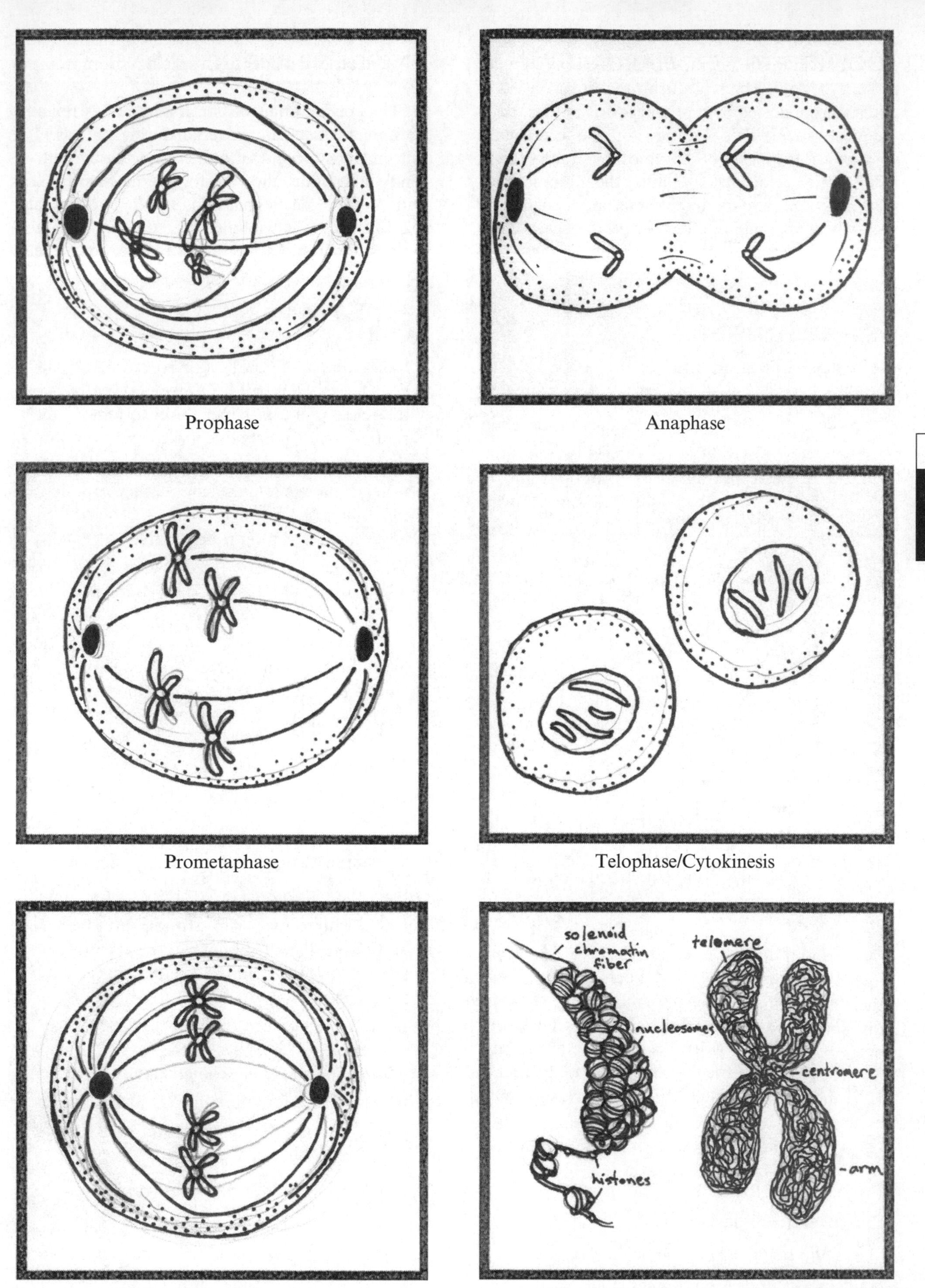
Prophase
Anaphase
Prometaphase
Telophase/Cytokinesis
Metaphase
solenoid chromatin fiber
telomere
nucleosomes
centromere
histones
arm
Chromatin/Chromosome Structure

packed structures that wrap the DNA onto a complex protein scaffold. The typical chromosome consists of two arms connected by a primary constriction site at the centromere. The location of the centromere determines the type of chromosome (acrocentric, metacentric, telocentric). Along the chromosome, there can be secondary and even tertiary constriction sites, which are regions of heavier protein scaffolding. The type of chromosome and location of these sites can often be used to identify a specific chromosome in a cell nucleus.

The condensation of chromatin allows for easy transfer of all genetic material during mitosis. The centromere is a point of attachment for the mitotic spindle fiber, thus allowing for the mechanical separation of the genetic material into the two daughter cells. Today, you will also be studying the morphology of chromosomes in Fava bean root tip cells.

PROCEDURES

Feulgen Reaction

- This protocol calls for the use of hot acid. Please be sure to do all work with acids in the hood, and be sure to wear all appropriate PPE.

1. Using forceps transfer one root tip from 70% ethanol to a beaker of distilled water. Always grab the root tips by the cut end. Incubate for 3 min.
2. Transfer the root tip to the beaker in the hood containing 1N HCl at 60°C. Rinse the forceps with distilled water. Incubate the root tips for 10 min.
3. Still in the hood, remove the acid from the hot plate and slowly decant the acid and root tips into a beaker with a larger volume of cooler distilled water.
4. Still in the hood, decant the acidic solution, keeping the roots in the beaker.
5. Still in the hood, add 20 mL of Schiff's reagent. Cover the beaker with Parafilm, and block all light from entering the beaker with a wrapping of aluminum foil. Incubate for 30 min.
6. Still in the hood, decant the Schiff's reagent, keeping the roots in the beaker. Add 20 mL of fresh bisulfate bleaching solution to the beaker (combine 180 mL of distilled water with 10 mL of 10% Na-bisulfite and 10 mL of 1N HCl). Incubate for 2 minutes, and then remove the liquid. Repeat this wash twice.
7. Still in the fume hood, rinse the root tips 3 times with distilled water. Keep the root tips in water after the last wash.

Squash Preparation

1. In the hood, label your microscope slide and add a drop of 45% acetic acid.
2. Still in the hood, grab your root tip from the cut end with a pair of forceps. Gently rub the root tip onto a paper towel to remove the root cap, then immediately place the root into the acetic acid. Incubate for 1 min.
3. Using a glass rod, make sure that the root tip is fully immersed in the acetic acid. Then gently lower a cover slip onto the sample.
4. Wick away the liquid from underneath the cover slip, using a paper towel. Be careful not to move the cover slip at all.
5. Using a pencil eraser, squash your root tip with direct, even pressure. The goal is to spread the cells of the root tip into an even, single layer. Continue to wick away excess liquid with a paper towel. Again, do not move the cover slip.

Permanent Preparation and Counterstaining

1. Place the slide onto a piece of dry ice for 5 min.
2. Pry off the cover slip with a razor blade and immediately plunge the slide into a Coplin jar filled with 95% ethanol. Incubate for 1 min.
3. Rinse the slide for 1 min with 95% ethanol, then immerse for 10 sec in the Coplin jar containing the fast green counterstain.
4. Rinse the slide for 1 min in 95% ethanol, twice.
5. Place a drop of 50% glycerol onto the sample, and cover with a new cover slip.
6. Examine the sample under the light microscope with the low-power and high-power objectives and make a note of the appearance on your datasheet. Collect the requested data for your datasheet.

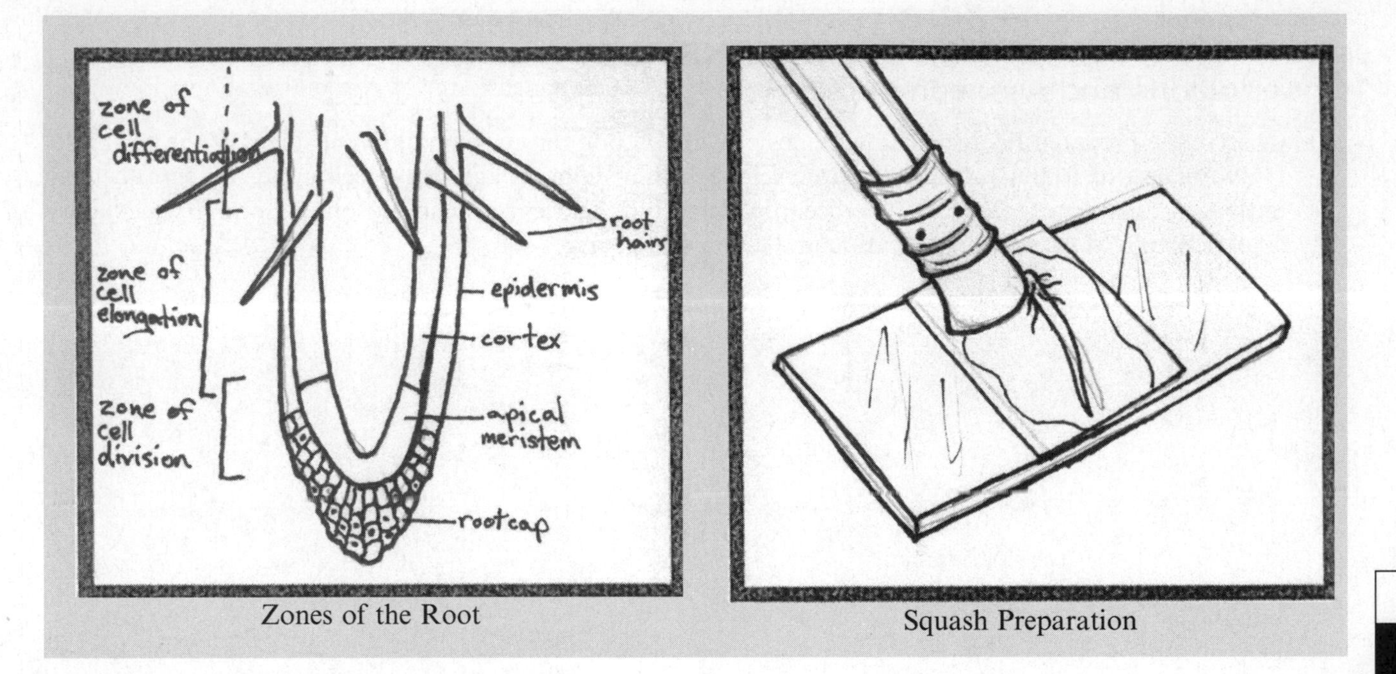

Zones of the Root

Squash Preparation

WORKSHEET

Observations and Assessments

1. Observe the prepared Feulgen-stained slides provided. Look for nuclei in different stages of the cell cycle. Draw images of four Feulgen-stained nuclei, each in a unique stage of the cell cycle. Be sure to label important structures (nucleoli, acrocentric/metacentric/telocentric chromosomes, primary/secondary constrictions). Mark the magnification and add a scale bar.

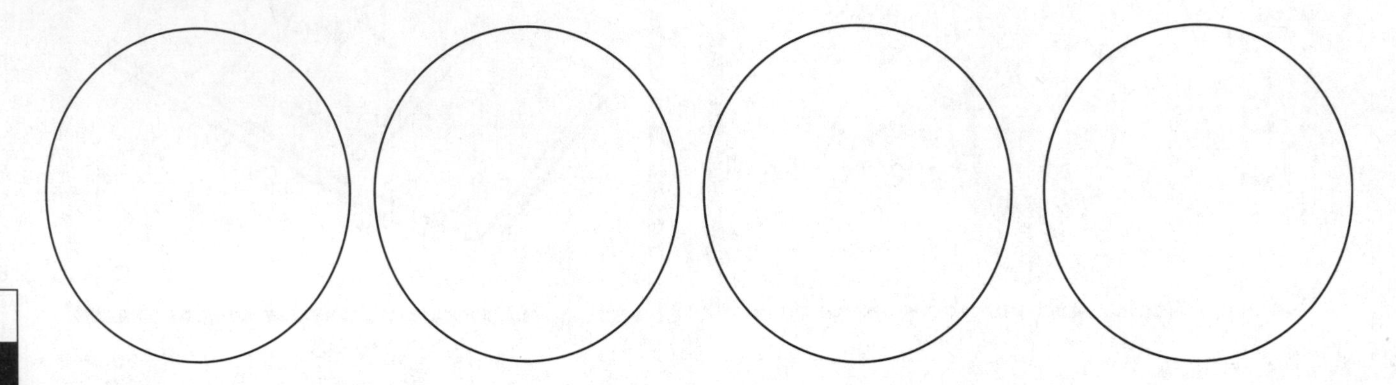

2. What do primary constriction sites on the chromosome represent? What do secondary constriction sites on the chromosome represent? Describe the Feulgen staining at these constriction sites. What chromosomal structures are altering the staining?

3. How many chromosomes do fava bean nuclei contain? What are the haploid (n) and diploid (2n) numbers for the fava bean? How many chromosomes per cell are acrocentric? How many are metacentric? How many are telocentric? Can you find any nuclei in your sample that contain a different number of chromosomes from the other cells? How can you explain any differences in chromosome number?

4. Determine and record the stage of cell division of 25 cells in the meristematic zone, the elongation zone, and the maturation zone. Record your findings in the table below. You will combine your data with the rest of the class.

Root Region	Interphase	Prophase	Prometaphase	Metaphase	Anaphase	Telophase
Meristematic						
Elongation						
Maturation						

5. Figure 14.1 in Karp (page 561) represents a pie chart of the phases of the cell cycle in a typical, mature cell. The area in the pie chart is representative of average time spent in each phase. There are *immature* cells in the root tip that are meristematic and actively dividing to form new cells. Below, draw a more representative cell cycle pie chart for these types of cells, using the data that you collected for question 4. You may use your data or the entire class's data, but please designate which data set you choose.

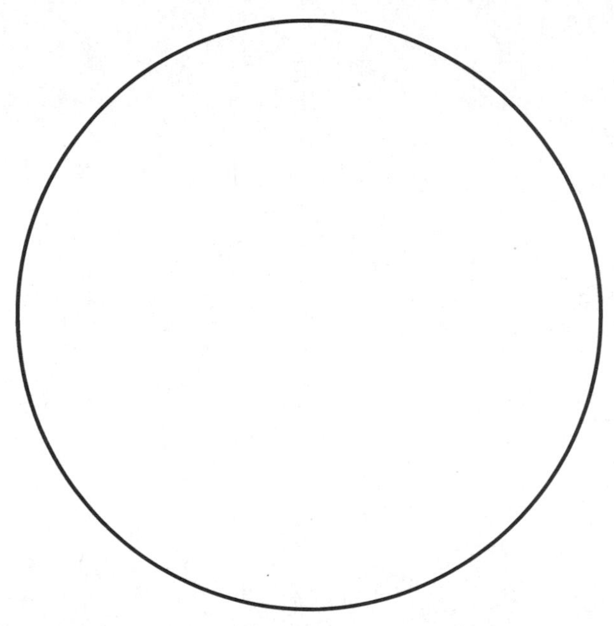

DISCUSSION

Discussion Questions

1. Discuss the targets of Feulgen staining in the chromosome. What does it stain? What parts of the chromosome are left unstained? Describe the appearance of both.
2. Discuss the differences between cell division and cell growth. Discuss how each is required for the expansion/production of new tissues. Discuss the points in an organism's development when either cell division or cell growth may be predominant.
3. Describe an experimental design that would determine the difference between cells in Gap phase and cells in S-phase. Use the guide to staining techniques in Appendix A for ideas. Your method should be quantitative.
4. Discuss the role of the cytoskeleton and cytoskeletal proteins in mitosis. Detail the involvement of these elements at each stage.
5. In this experiment, we studied mitotic cell division and the cell cycle. Describe an experimental design that would observe meiosis in a plant cell. What tissue source would you choose? Would you use the same staining method, or choose a different one? Discuss your expected results.

Staining for Cell Viability and Apoptosis

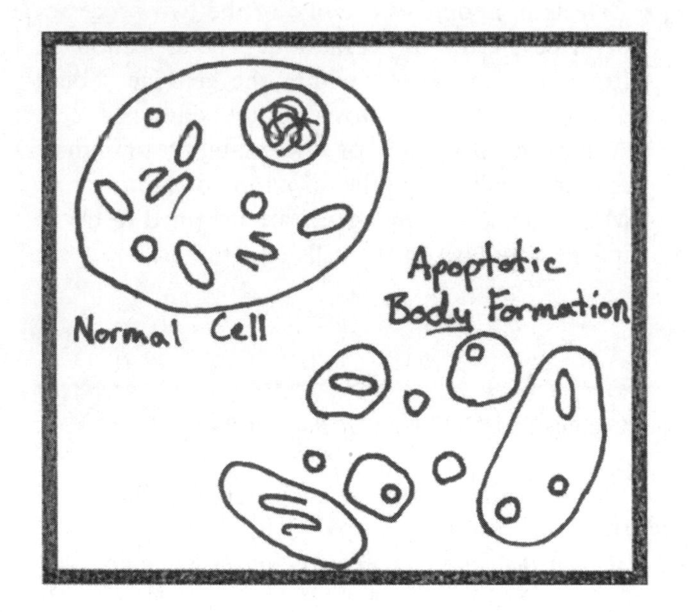

Objectives

- Treat a cell sample to induce apoptosis,
- Use two different microscopy-based staining methods to differentiate between living and dead cells.

15
OVERVIEW

DESCRIPTION

There are two main types of cell death: uncontrolled necrosis and programmed cell death (PCD). PCD is a tightly regulated process that is induced to safely break down a cell. There are several causes of PCD, which include irreparable DNA damage, viral infection, aging and cell turnover, tissue sculpting. Caspases are the activators of apoptosis. There are several ways to study each of the types of cell death: the presence of dead or lysed cells, genetic or antibody-based screening, and the detection of DNA. Today, you will be detecting living and dead cells using two fluorescent dyes and one colorimetric stain. By using the fluorescent dyes: fluorescein diacetate (FDA) and propidium iodide (PI) together, you can detect living and dead cells, respectively. Trypan blue is another simple staining procedure that stains dead cells blue. Using these methods you will look at the induction of apoptosis and its effects on the cell culture.

CONCEPTS & VOCABULARY

- Apoptosis
- CAD
- Caspase
- FDA
- Ionizing radiation
- Lamin
- Necrosis
- Oxidative stress
- Trypan blue
- Programmed cell death
- Propidium iodide
- Senescence
- Telomere

CURRENT APPLICATIONS

- Cell culture viability assays are the first test for all new pharmaceuticals. If the drug is particularly toxic, it will induce apoptosis or necrosis in a high percentage of cells in a suspension culture. Fluorescent-labeling with FDA and PI coupled with flow cytometry, which counts the number of cells in a sample with each fluorescent signal, can determine the toxicity of a compound within 10 minutes.
- Defects in apoptosis are one of the two processes involved in tumor formation. The detection of irreparable mutations within the genome should always lead to apoptosis of the damaged cell. A lack of detection or the inability to induce apoptosis will lead to the formation of a tumor cell. Many chemotherapy agents are designed to try to induce apoptosis in the cells of a tumor.

REFERENCES

Harris, JR, Graham, J, Rickwood, D. (2006). Quantitation of cell counts and viability. *Cell Biology Protocols*. (67–70). Hoboken, NJ: John Wiley & Sons, Inc.

Jones KH, Sent, JA. (1984). An improved method to determine cell viability by simultaneous staining with fluorescein diacetatepropidium iodide. *J Histochemistry and Cytochemistry*. 33:77–79.

Karp, G. (2010). Apoptosis (Programmed Cell Death). *Cell and Molecular Biology: Concepts and Experiments*. (642–646). Hoboken, NJ: John Wiley & Sons, Inc.

BACKGROUND

TYPES AND CAUSES OF CELL DEATH

There is no such thing as a normal, immortal cell. All cells have a limited number of divisions or a limited lifespan. While there are some cells that are extremely long-lived, eventually they will be turned over by cell death and replaced by new cell division.

There are two main types of cell death. The first is called necrosis, which is uncontrolled and caused by extreme damage or distress. When a cell is exposed to extreme heat or cold, very low or high pH, or direct physical damage, it will immediately begin to break down and potentially release is components. This uncontrolled breakdown not only destroys the cell itself, but the release of various enzymes and toxic intermediates can also cause damage to other cells in the surrounding tissue. Programmed cell death (PCD) is a tightly regulated process that can safely break apart a cell, reducing risks of damaged to the surrounding damaged tissue. PCD may be apoptotic or autophagic. Generally apoptosis is induced to rapidly break down a cell, whereas autophagic PCD is usually associated with aging

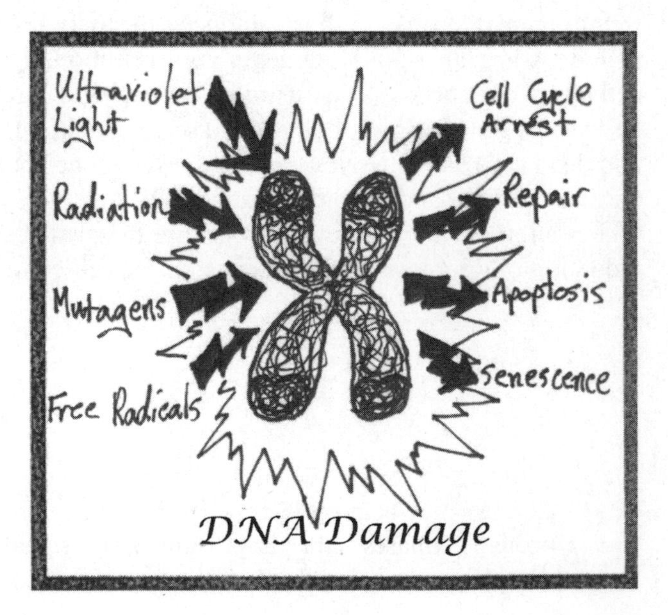

(senescence). There are several causes of PCD, which include irreparable DNA damage, viral infection, aging and cell turnover, and tissue sculpting. Each of these causes has different molecular triggers, such as DNA damage. The source of this trigger dictates the type of PCD. A cell is directed to begin to engage in apoptosis after activation by a T-lymphocyte. T-cells recognize one of the triggers of apoptosis and induce the production of cysteine proteases, called caspases.

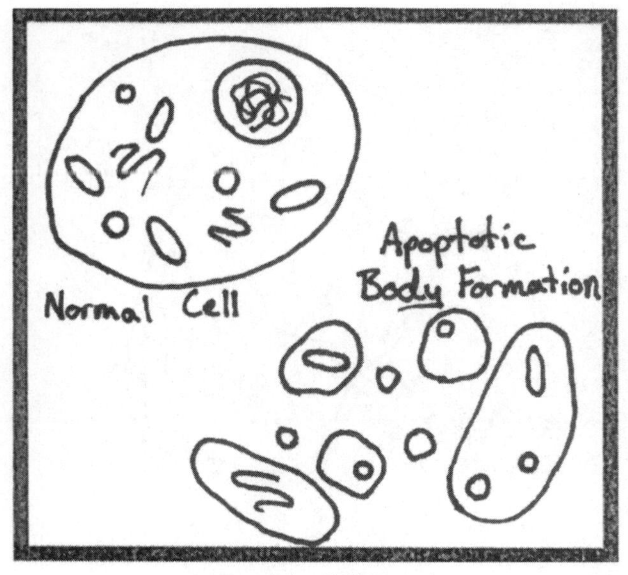

Apoptotic Cell Death

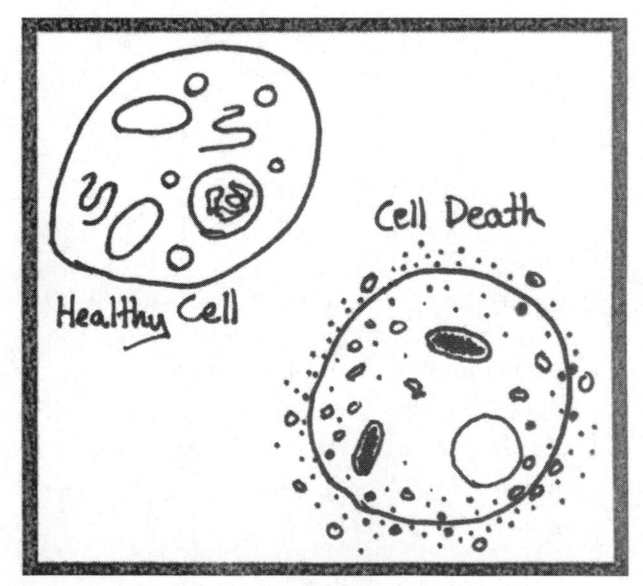

Autophagic Cell Death

INDUCTION OF APOPTOSIS

Caspases are the activators of apoptosis. They modify other inactive apoptosis proteins and initiate a cascade that begins the process. For an example, protein kinases are important cell signal molecules, lamins are involved in the breakdown of nucleases, Caspase-Activated DNase (CAD) degrades genomic DNA, and caspases activate directly to break down cytoskeletal proteins to destabilize and alter the structure of the cell.

Apoptosis induced by an external trigger is called extrinsic induction. External triggers include ionizing radiation, increases in temperature, the presence of viruses, and the presence of toxins. Extrinsic induction is triggered by a release of a tumor necrosis factor (TNF), and is regulated by a caspase-8 cascade. Apoptosis caused by an internal trigger is called intrinsic induction. Internal triggers include DNA damage, hypoxia, increased Ca + + levels, viral infection, and oxidative distress. Intrinsic induction is induced by a Bcl-2 family cascade, generally the Bax protein found on the mitochondrial membrane. Intrinsic induction is regulated by caspase-9. Bax causes the release of cytochrome C from the mitochondria, which leads to the formation of the structure called the apoptosome. This complex regulates apoptosis after its formation.

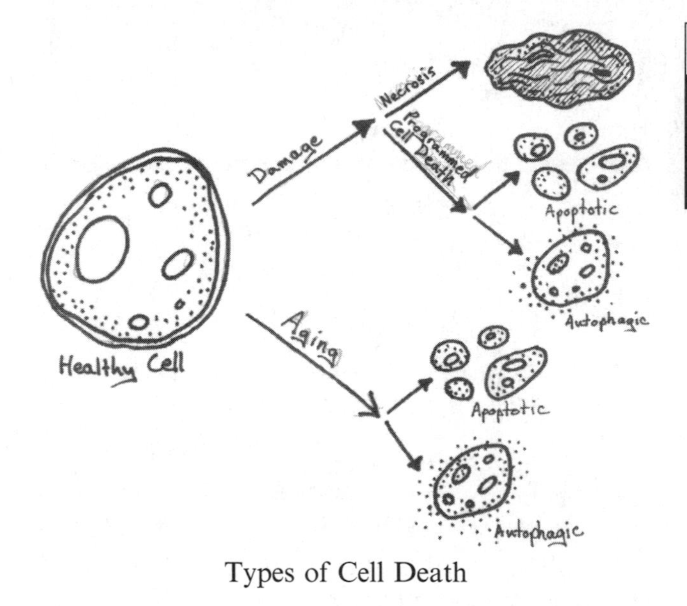

Types of Cell Death

DETECTION AND MEASUREMENT OF CELL DEATH

There are several ways to study each of the types of cell death. Necrosis can be easily identified by the presence of dead or lysed cells, or the detection of cytoplasmic components in the extracellular fluids. Apoptosis can be detected by genetic or antibody-based screening. Other common apoptosis assays include detection of phosphotidylserine translocation, caspase activation, histone phosphorylation, and DNA fragmentation. DNA damage can be easily screened for, using fluorescence, microscopy, or flow cell cytometry. Common fluorescent stains used to detect DNA damage include acridine orange mixed with ethidium bromide, or DAPI alone. Senescence can be distinguished from apoptosis. Genetic screens for senescence-related genes is a simple and common assay. Other screens include observation in telomere-shortening, detection of autophagy, and DNA damage linked to oxidative stress. A simple oxidative stress assay is the screen using the fluorescent stain dichlorofluorescein diacetate (DCFDA).

Today you will be detecting living and dead cells using two fluorescent dyes and one colorimetric stain. By using the fluorescent dyes fluorescein diacetate (FDA) and propidium iodide (PI) together, you can detect living and dead cells, respectively. Trypan blue is another simple staining procedure that stains dead cells blue. Using these methods, you will look at the induction of apoptosis and its effects on the cell culture.

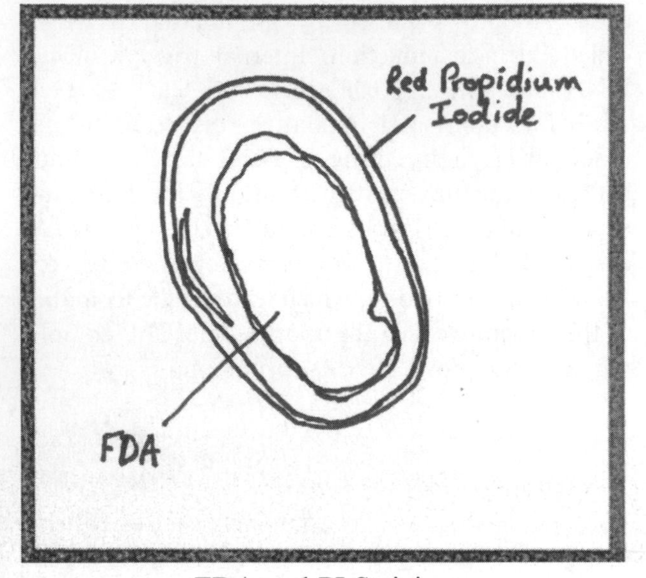

FDA and PI Staining

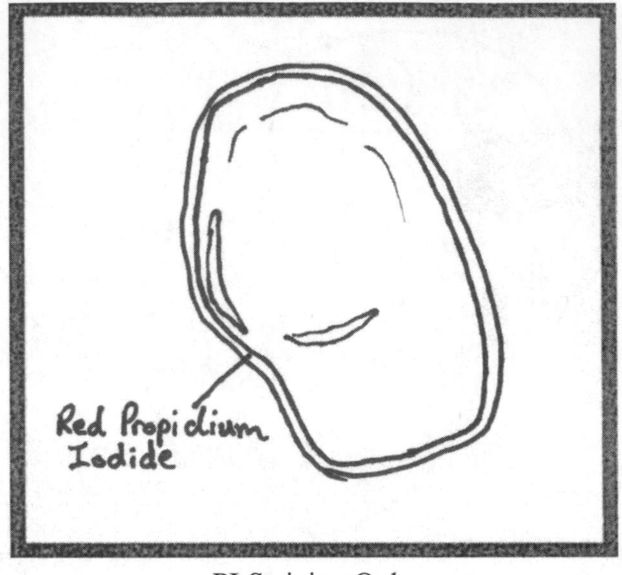

PI Staining Only

15

OVERVIEW

PROCEDURES

Induction of Necrosis and Apoptosis in a Mammalian Cell Culture

1. Obtain two mammalian cell cultures from your instructor. One will be marked Untreated, and one will be marked Oxidative.
2. Transfer 1 mL of the Untreated culture into 2 wells of a 6-well plate.
3. Transfer 1 mL of the Oxidative culture into a different well of the same 6-well plate.
4. Add an additional 2 mL of fresh culture medium to each of the 3 wells.
5. Label the wells Control, Heat, and Oxidative Stress.
6. Transfer the culture from the Heat well into a screw-cup, 5 mL tube. Seal the tube, and heat it at 70°C for 10 minutes. Then, return it to the well.
7. Test each sample with the two staining methods below at 0 minutes, 30 minutes, and 60 minutes.

FDA/PI-Staining Procedure

1. Transfer 1 mL of each treatment sample to a clean tube. Label the tubes Control, Heat, Oxidative.
2. Centrifuge each cell suspension at 1000 rpm for 5 minutes. Remove the supernatant, and resuspend in 3 mL of fresh medium.
3. Add 3 μL of FDA staining solution and 3 μL PI staining solution to the Control sample cells.
4. Incubate at room temperature for 3 minutes, then place on ice.
5. Load a drop of the cell suspension onto a Neubauer haemocytometer. Add a coverslip.
6. Observe the sample, using the fluorescence microscope using FITC and TRITC filter cubes. Start counting the numbers of green and red cells in 25 random squares on the grid.
7. Clean off the haemocytometer, and repeat steps 1–5 for the Heat and Oxidative samples.

8. For the 30-minute and 60-minute samples, start with a fresh 1 mL transfer from each culture.

Trypan Blue-Staining Procedure

1. Transfer 1 mL of each treatment sample to a clean tube. Label the tubes Control, Heat, Oxidative.
2. Centrifuge each cell suspension at 1000 rpm for 5 minutes. Remove the supernatant, and resuspend in 250 μL of fresh medium.
3. Add 250 μL of Trypan blue staining solution to the Control sample, and mix by inversion.
4. Immediately load a drop of the cell suspension onto a Neubauer haemocytometer. Add a coverslip.
5. Observe the sample using the light microscope. Start counting the numbers of clear and blue cells in 25 random squares on the grid.
6. Clean off the haemocytometer and repeat steps 1–5 for the Heat and Oxidative samples.
7. For the 30-minute and 60-minute samples, start with a fresh 1 mL transfer from each culture.

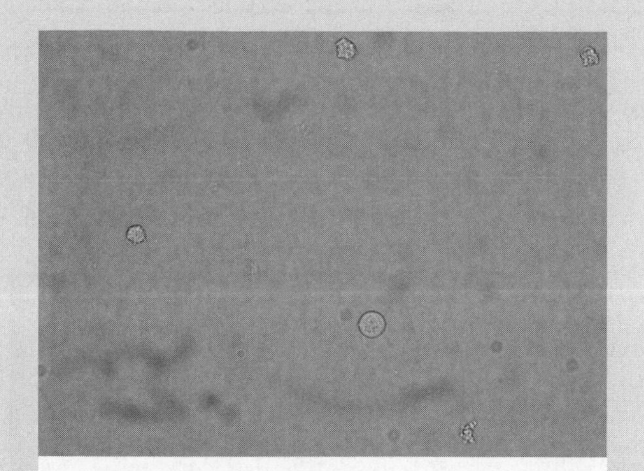

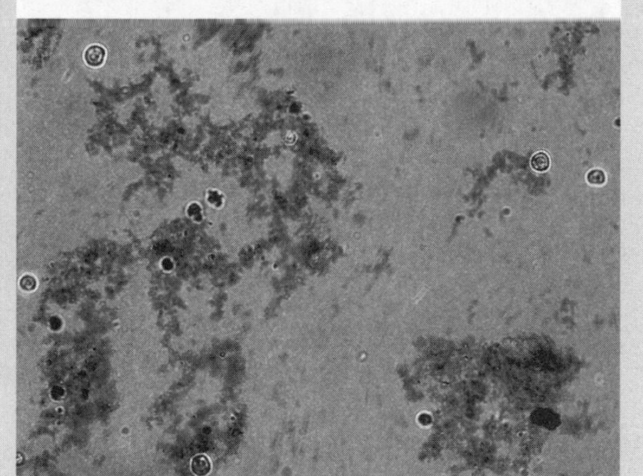

Trypan Blue Staining of CHO Cells

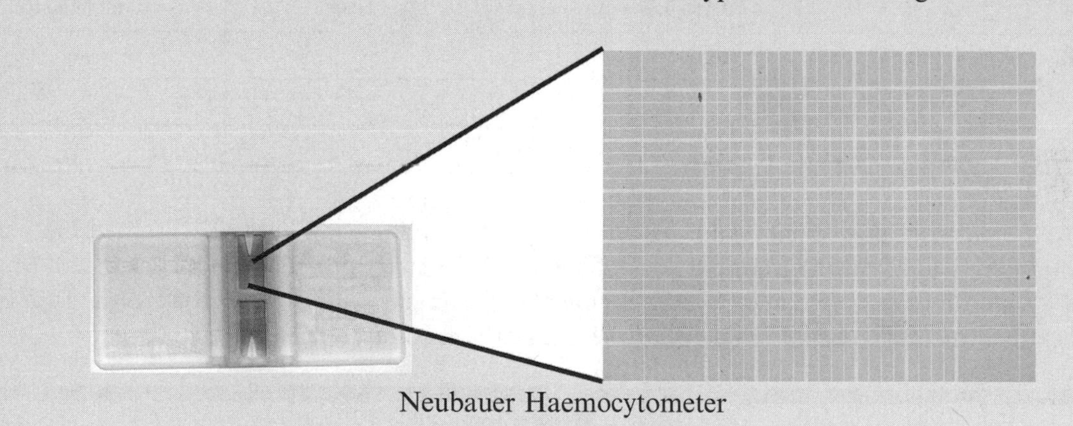

Neubauer Haemocytometer

15

OVERVIEW

WORKSHEET

1. Observe your FDA/PI stained and unstained control samples, using epifluorescence. How does the stained sample look? Draw an example sketch for both in the space provided below. Using two colors, sketch a drawing with both FDA and PI staining. Remember to label all microscope images with the following: organism, cell type, magnification, and stain; add a scale bar at the bottom right of each sketch.

15 OVERVIEW

2. Record your cell counts for the FDA/PI-stained samples.

Sample	**0-Minutes**	**30-Minutes**	**60-Minutes**
Control			
Boiled			
H_2O_2-treated			

3. Observe your trypan blue stained and unstained control samples using light microscopy. How does the stained sample look? Draw an example sketch for both in the space provided below. Using color, sketch a drawing of the staining. Remember to label all microscope images with the following: organism, cell type, magnification, and stain; add a scale bar at the bottom right of each sketch.

4. Record your cell counts for the trypan blue-stained samples.

Sample	0-Minutes	30-Minutes	60-Minutes
Control			
Boiled			
H_2O_2-treated			

5. What type of cell death is caused by the boiling treatment? Do your staining results confirm this? How do your FDA/PI-results compare with your trypan blue results? What type of cell death is caused by the hydrogen peroxide treatment? Do your staining results confirm this? How do your FDA/PI-results compare with your trypan blue results? Which method do you feel is more reliable?

15

OVERVIEW

DISCUSSION

1. Discuss the differences between necrosis and apoptosis. Discuss the differences between intrinsic and extrinsic induction of apoptosis.
2. Research the staining targets for fluorescein diacetate and propidium iodide. Discuss how these stains differentiate between living and dead cells, respectively.
3. Research the staining target for trypan blue staining. Discuss how this staining method differentiates between living and dead cells.
4. Discuss how difficult it was to hand-count your cell samples for each time point. Research flow cytometry, and discuss the importance of one of these devices in a cell biology laboratory. Discuss other possible assays that can be performed with a flow cytometry machine.
5. Choose an organism, and find three genes that increase in expression in cells that are senescing. Discuss the experimental design for a genetic screen for this organism.

Staining of Immune Cells

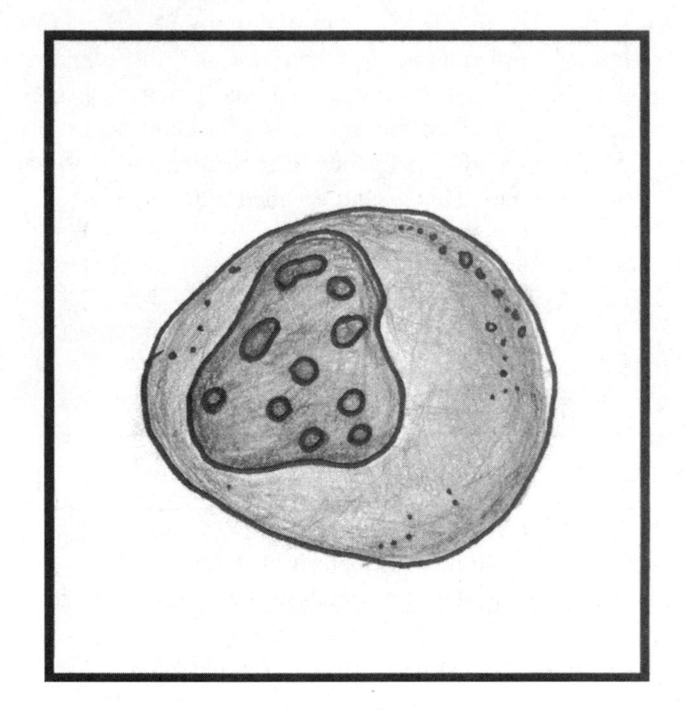

Objectives

- To make a Wright–Giemsa-stained blood smear
- To observe and catalogue the various types of blood cells

16 OVERVIEW

DESCRIPTION

There are many types of blood cells found in mammalian blood. Erythrocytes, or red blood cells, are essential for proper respiration. These cells carry oxygen into the tissues and carry carbon dioxide away from the tissues. Platelets are essential for proper clotting and release a number of chemicals important for maintaining homeostasis. The remaining cells are called white blood cells. These cells are part of either the innate or the adaptive immune system. The cells of the innate immune system engulf, destroy, or present foreign proteins called antigens. The cells of the adaptive immune system will also destroy foreign antigens, but they play a role in creating the cellular memory of these antigens as well. If the antigen is recognized by the immune system a second time, the secondary immune response will be faster and stronger. Blood smears can be labeled by the Wright-Giemsa staining method. This procedure uses a mixture of acidophilic and basophilic stains to label the nuclei, cytoplasm, and intra-cellular granules of the immune cells. Each of the different types of immune cells will have their own characteristic staining pattern and morphology. The ability to discern the population of immune cells within a blood sample is an invaluable tool to be used in the diagnosis of disease.

CONCEPTS & VOCABULARY

- Antibody/immunoglobin
- Antigen/epitope
- B-lymphocyte (B-cell)/plasma cell
- Basophil
- Eosinophil
- Memory cell
- Monocyte/macrophage
- Natural killer cell
- Neutrophil
- T-lymphocyte

CURRENT APPLICATIONS

- Occasionally the immune system will become confused and recognize an antigen on something that is not an infectious microorganism or tumor cell. This may be a protein on a pollen grain, a peptide within a nut protein, or the venom from a bee sting. Once this antigen has been recognized by the immune system, an inflammation response will always be induced near the point of contact. This is the cause of allergies.
- Very early in development the immune cells are trained to distinguish *self* from *non-self*, in order to prevent any auto-immune reactions. Later in development, specialized immune cells in the intestines train our immune system further to recognize and ignore food antigens. This system is called oral tolerance.

REFERENCE

Karp, G. (2010). The immune response: Uses of antibodies. *Cell and Molecular Biology: Concepts and Experiments* (682–709; 763–765). Hoboken, NJ: John Wiley & Sons, Inc.

BACKGROUND

TYPES OF IMMUNITY & CELLS OF THE IMMUNE SYSTEM

The immune system provides defense against infectious organisms and viruses, tumor cell formation, and foreign compounds. Every immune response begins with recognition of a foreign antigen by specific cells of the immune system. This is followed by inflammation of the site and release of hormones and chemotactic compounds that attract other immune cells to the area. The cells of the immune system are found in the blood, in the lymphatic fluid, and in the lymph nodes. All blood cells are produced in the bone marrow. Hematopoietic stem cells differentiate into either myeloid progenitor cells or lymphoid progenitor cells. Lymphoid progenitor cells produce T- and B-lymphocytes and natural killer cells. The myeloid progenitor cells produce all of the other blood cells, including: erythrocytes, platelets, the eosinophils, neutrophils, basophils, mast cells, dendritic cells, and monocytes – which mature into macrophages.

There are two types of immunity: innate and adaptive. Innate immunity is provided by cells that are always active. These cells include the phagocytes, natural killer cells, and granulocytes. Macrophages sweep through the blood stream looking for foreign antigens and engulf them. The complement system produces pores in foreign membranes to weaken or destroy infectious cells. Natural killer (NK) cells induce apoptosis in tumor cells and cells infected with viruses. Interferon (IFN) is released by infected cells as a signal to surrounding cells to prepare for infection. All of these systems are typically non-specific and always active. Granulocytes like basophils, eosinophils, and neutrophils contain granular vesicles full of compounds important to the innate immune response. These compounds may be digestive enzymes, used to break down foreign cells and proteins. They may also be cell-signaling molecules that cause inflammation and attract the other cells of the immune system. All of the cells in the innate immune system act individually to respond to the immediate threat of an infection or tumor cell formation.

Adaptive immunity can be broken down to two subtypes: the humoral response and the cell-mediated response. Adaptive immunity involves immune cells called lymphocytes.

The humoral response involves B-cells, which are responsible for the production of antibodies in the body. Antibodies, or immunoglobins, are large protein complexes, which bind to foreign antigens. Each B-cell only produces one type of antibody, which has an antigen-binding region that recognizes a specific part of the antigen known as an epitope. There may be several antibodies that can combine to several different epitopes of an antigen. But every B-cell in the body is specific to one epitope. When the antibodies bound to the exterior of the plasma membrane come in contact with their specific antigen, the B-cell matures into a plasma cell. Plasma cells are capable of producing large quantities of antibodies, which are released as free-floating protein complexes into the bloodstream. These free antibodies will then bind to the antigen wherever they find it. The antigen/antibody complex can signal other immune cells to attack and destroy the foreign antigen, or the antibodies are capable of forming large aggregates with the antigen and preventing its free movements though out the organism.

CELLS OF THE IMMUNE SYSTEM

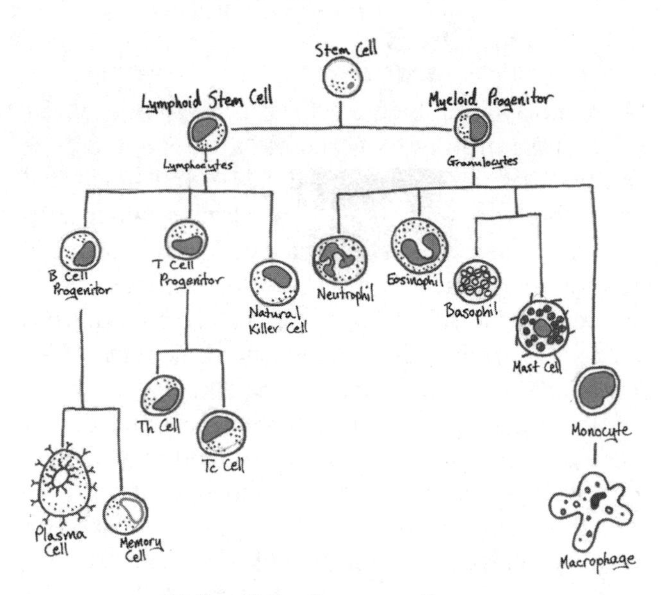

Cells of the Immune System

Thrombocytes

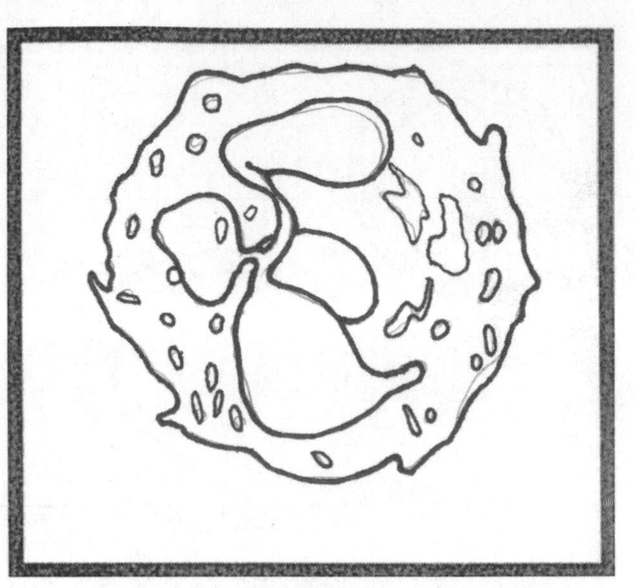
Neutrophils

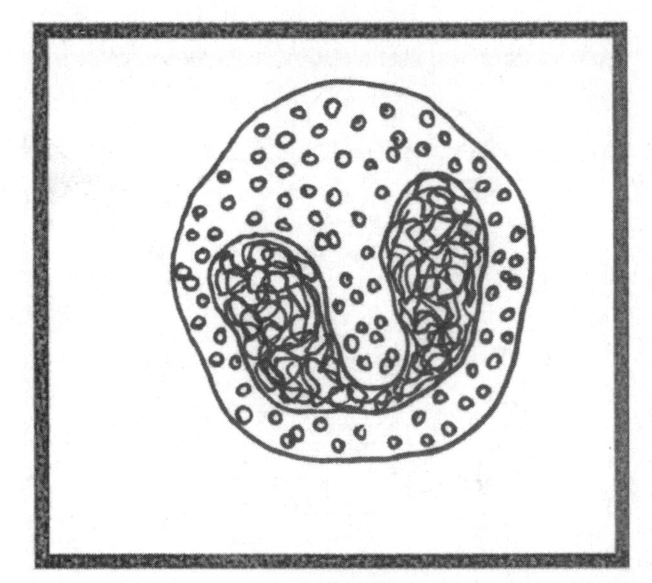
Eosinophils

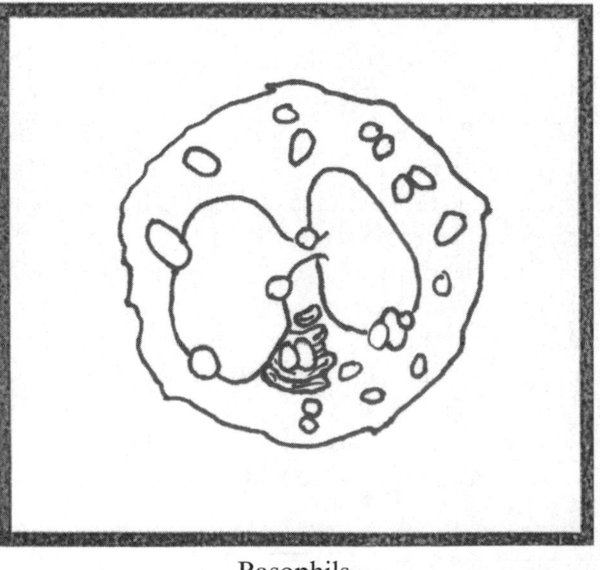
Basophils

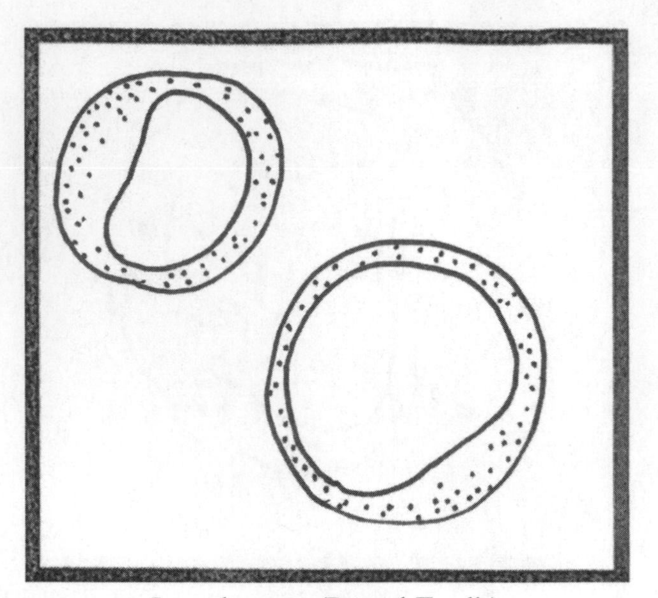

Lymphocytes (B- and T-cells)

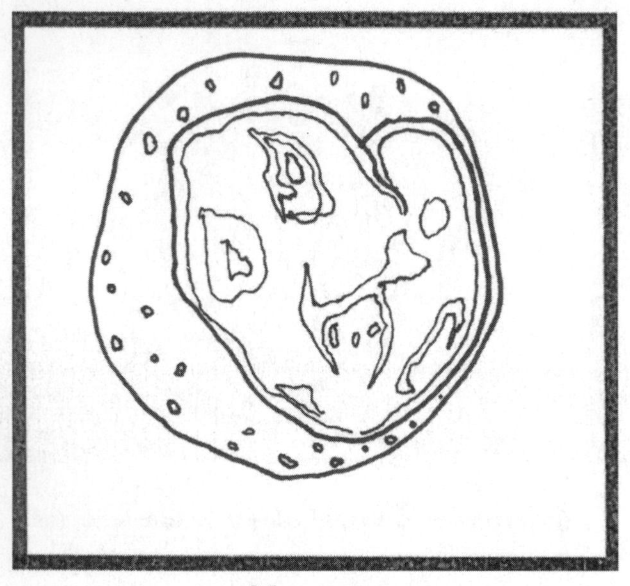

Monocytes

The cell-mediated response involves T-lymphocytes, or T-cells. There are three types of T-cells: T-helper (Th) cells, cytotoxic T-cells (Tc), and T-regulatory cells (Treg). T-helper cells are the main activator of the immune system. They recognize foreign antigens presented by other immune cells. For example, macrophages process the antigens that they engulf and present them on the exterior of their plasma membrane. Once a Th cell finds a bound antigen, it will activate both a stronger innate immune response as well as the adaptive immune response. T-helper cells can activate other T-cells and B-cells. If a cell is infected with a virus, or expressing proteins that identify it to be a tumor cell, the T-helper cell will activate cytotoxic T-cells and Natural Killer cells. Tc cells induce apoptosis in infected cells, which leads to cell death. Tc cells are a frontline defense against viral infection and the prevention of cancer. Regulatory T-cells are not well studied, but they are known to be involved with the up-regulation or down-regulation of the cell-mediated immunity. The Human Immunodeficiency Virus (HIV) produces Acquired Immune Deficiency Syndrome (AIDS) by infecting and destroying Th cells and macrophages. By depleting these two cell types in the bloodstream and lymphatic system, HIV effectively cripples the immune system.

WRIGHT-GIEMSA STAINING OF BLOOD CELLS

A monolayer of blood cells can be produced by smearing a drop of blood across a glass microscope slide. Observation of an unstained blood smear will provide little information, as most white blood cells are clear and red blood cells are very homogenous in morphology. The Wright–Giemsa staining method was devised to selectively stain for white blood cells using the acidophilic methylene blue stain, the basophilic methylene azure stain, and the eosinates of both stains. These four compounds will selectively stain acid or basic structures in the blood cell. Today you will create a blood smear and stain it with the Wright-Giemsa method. Then you will observe, identify, and sketch each type of immune cell.

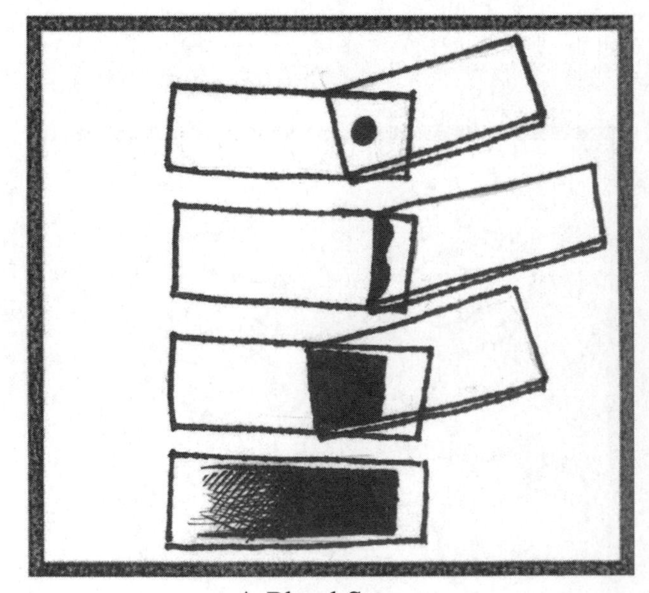

A Blood Smear

PROCEDURES

Blood Smear

1. Place a 100–200 μL drop onto a clean slide.
2. Streak thin smears across the slide by means of a second clean slide.
3. Air-dry quickly by carefully waving the slide in the air.

Staining

1. Cover the entire surface of the smear with Wright-Giemsa Stain (about 1.0 mL). Incubate for 5 minutes.
2. Remove stain and wash with 2.0 mL of Phosphate Buffer, pH 6. Incubate 10 minutes
3. Rinse stained smear with dd-water until the edges turn light red.
4. Blot or air dry.
5. Add large coverslip, and observe using light microscope. Observe all of the stained white blood cells. Take the time to find erythrocytes and platelets in your sample as well.

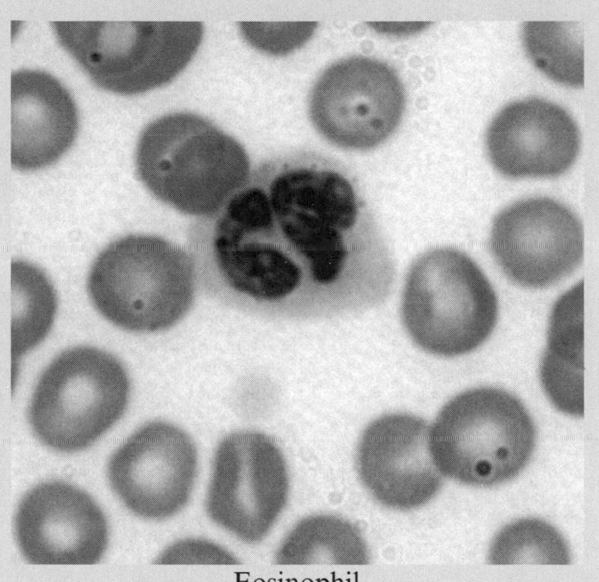

Eosinophil

Lymphocyte

Basophil

Monocyte

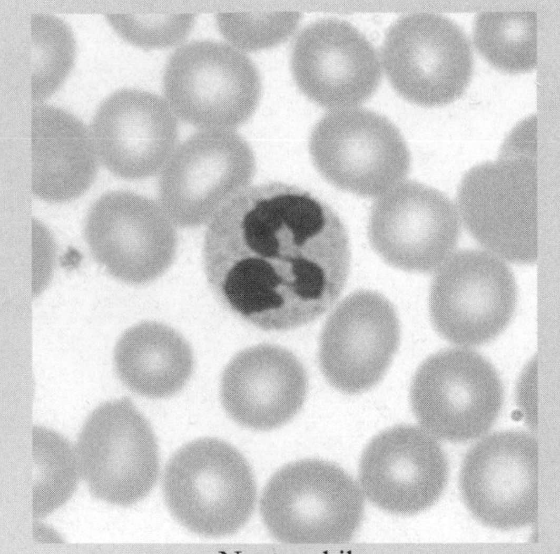
Neutrophil

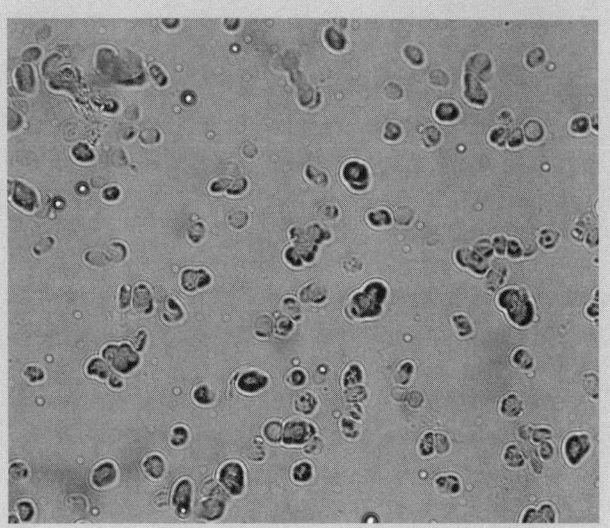
Erythrocyte and Platelets

WORKSHEET

1. Draw examples of the monocytes/macrophages and lymphocytes that your observe on your blood smear. Describe any visible structures, and draw an example sketch in the space provided below. Remember to label all microscope images with the following: organism, cell type, magnification, and stain; add a scale bar at the bottom right of each sketch.

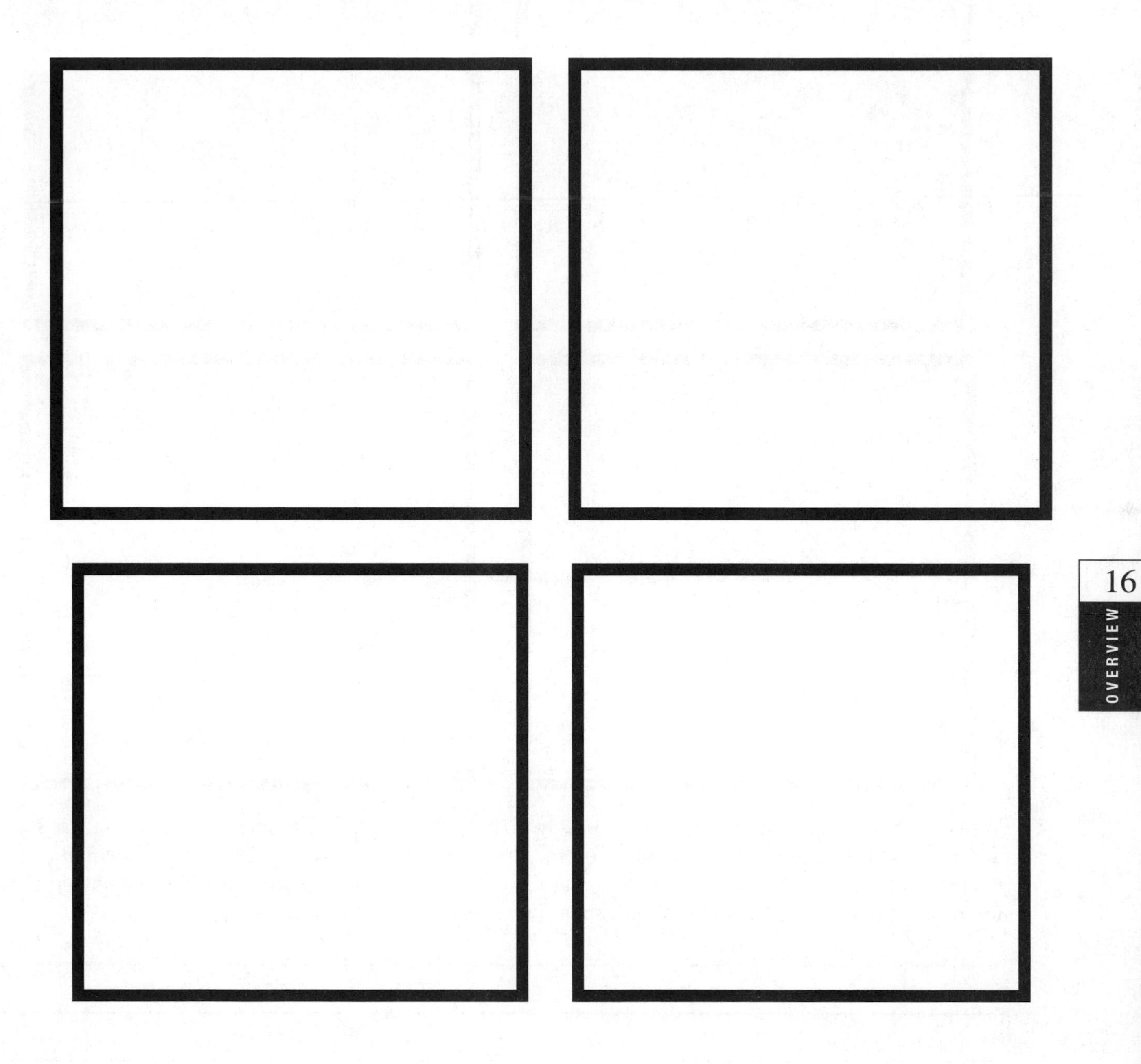

2. Draw examples of the granulocytes that you observe on your blood smear. Describe any visible structures, and draw an example sketch in the space provided below. Remember to label all microscope images with the following: organism, cell type, magnification, and stain; add a scale bar at the bottom right of each sketch.

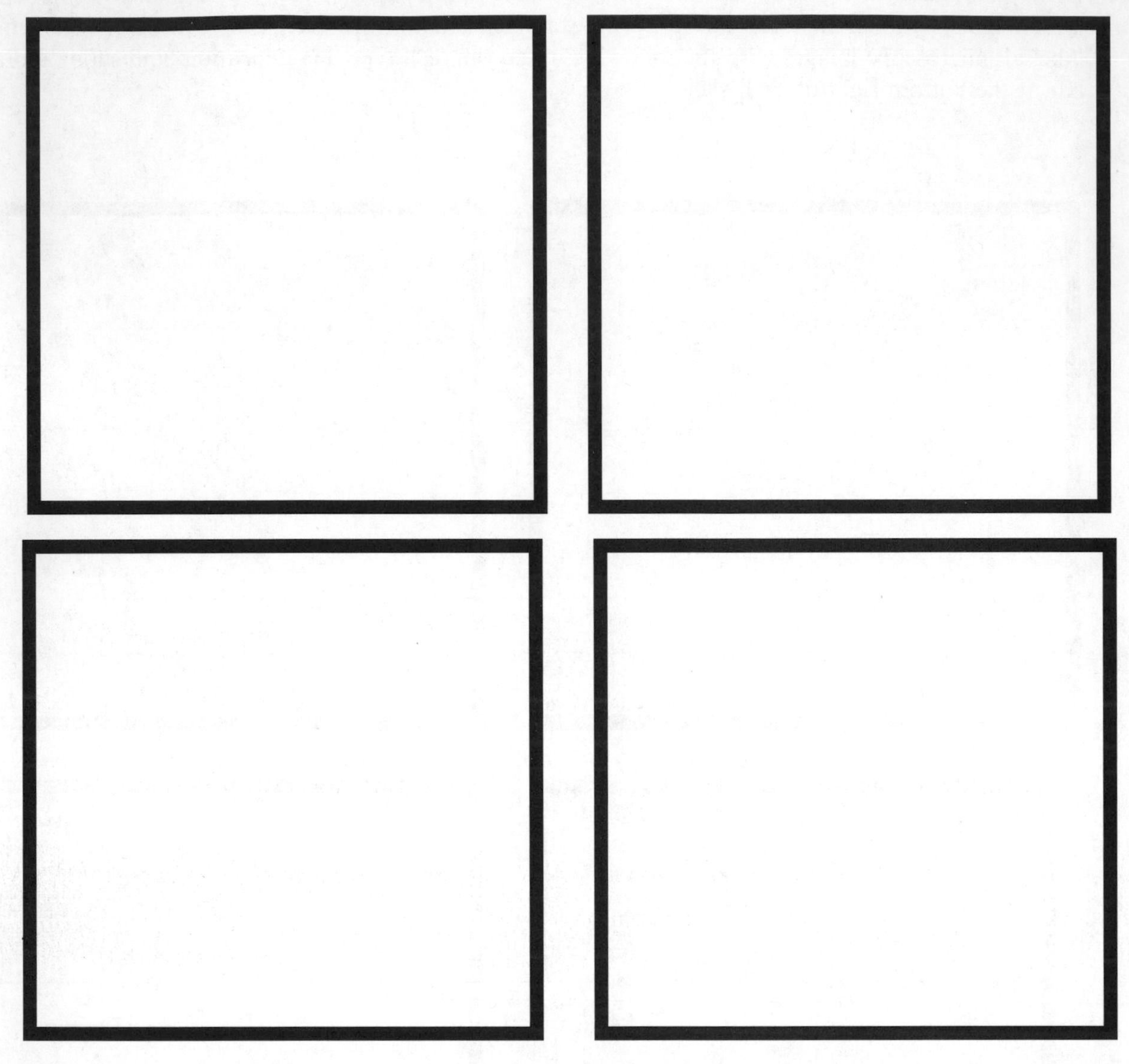

3. Wright-Giemsa stain contains the acidophilic methylene blue stain, the basophilic methylene azure stain, and the eosinates of both stains. These two procedures can be separated into just Wright staining or just Giemsa staining. Using the table in the background section, discuss the differences in the staining profile for each type of cell if you did a Wright stain. Do the same for a Giemsa stain.

____

____

____

____

____

____

____

____

4. Blood smears are not typically used in medical analysis of blood samples. Typically, the blood cell count is measured using an automated cell sorter. Do some research in the library or on the Internet and describe how a complete blood cell count machine works.

5. There are several times when blood smear tests are still used for medical diagnosis, however. Again, do some research in the library or on the Internet, and describe one example of a medical test that still uses a blood smear and microscopic examination.

16

OVERVIEW

DISCUSSION

1. What type of white blood cell would you expect to see increased in a blood sample from a patient with a viral infection? What type of white blood cell would you expect to see increased in a blood sample from a patient with a bacterial infection? Explain your response.
2. You have a patient with a very high white-blood-cell count. How would you diagnose this patient? What further testing would you suggest?
3. You find a high concentration of basophils in a patient's blood sample. What other symptoms would you expect? Explain your response.
4. You find a low concentration of lymphocytes in a patient's blood sample. What other symptoms would you expect? Explain your response.
5. What did you notice about the amount of red blood cells and platelets? Discuss the functions of each cell type. Why are these cells more prevalent?

Protein Isolation, Fractionation, and SDS-PAGE

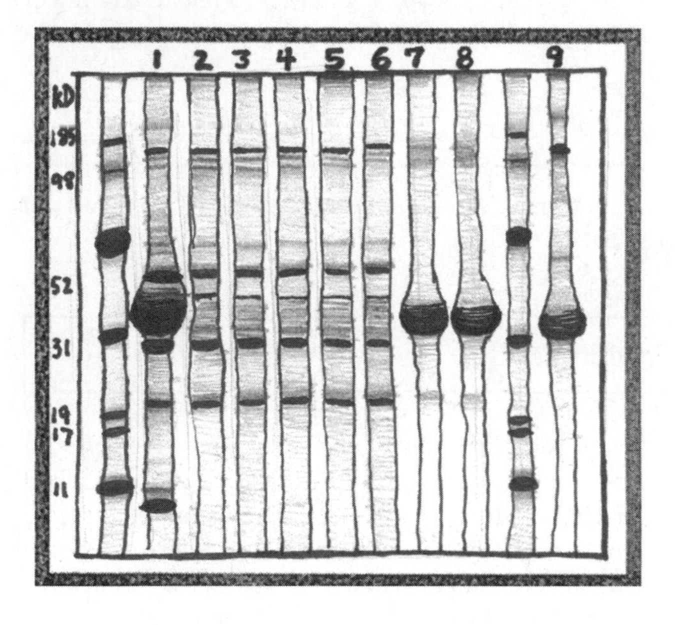

Objectives

- Detecting differences in identity between biological samples using protein analysis techniques.
- Isolating and separating proteins by size using solubility and gel electro phoresis.

DESCRIPTION

In most cases, proteins are the end product of gene expression. Proteins can be structural, involved in cell signaling or ligand binding, or enzymes. Within a species, the majority of structural proteins and enzymes will have identical sequence, size, and structure. However, in different species, there will be an increasing number of sequence and structural differences, which can be detected using a number of readily available molecular techniques. SDS-polyacrilamide gel electrophoresis (SDS-PAGE) is a technique that can be used to separate crude protein samples by mass. SDS is used to denature every protein in a sample, and the denatured sample is run on a polyacrilamide gel to separate each individual kind of protein. The gel is then stained, and specific protein bands can be visualized. SDS-PAGE gels can also be used to create Western blots, which use antibodies to detect specific proteins based on structural elements that are unique to each protein, called *antigens*. Today, you will be preparing crude protein extracts, fractionating them into soluble and insoluble proteins, and separating them on an SDS-PAGE gel to determine the identity of each sample.

CONCEPTS & VOCABULARY

- Antigen
- Denature/renature
- Immunodetection
- Primary antibody
- SDS-PAGE
- Secondary antibody
- Soluble/insoluble
- Western blotting

CURRENT APPLICATIONS

- Proteomic analysis is one of many of the new forms of robust molecular analysis available to biologists (genomics, transcriptomics, proteomics, metabolomics). Using 2-D gel electrophoresis, LC-MS-MS analysis, or protein microarrays, a researcher has the potential to observe the presence and concentration of all proteins in a biological sample. Coupled with powerful computer-based analysis, these techniques are a gateway to understanding of the protein nature of the proteome.

REFERENCE

Karp, G. (2010). Isolation, purification and fractionation of proteins. *Cell and Molecular Biology: Concepts and Experiments.* (734–740). Hoboken, NJ: John Wiley & Sons, Inc.

BACKGROUND

PROTEINS

Proteins are composed of amino acid monomers connected by a peptide linkage. They include peptides and their polymers. Proteins can exist as simple peptides, but they can also exist as large complexes of many peptides. Within the cell there are thousands of proteins. Some of them are structural, serving some mechanical function. Many of them are enzymes, controlling metabolic reactions or serving some sort of regulatory function. Beyond that, there are molecular pumps, protein motors, and even proteins involved in energy capture or transmission. Thousands of proteins are active at any given time, just to maintain the housekeeping functions of the cell. Other proteins are inserted into cell membranes or even secreted by the cell into the extracellular space.

PROTEIN SPECIFICITY

The amino acid polymers that make up peptides fold back upon themselves due to molecular interactions and the formation of ionic bonds. Later, covalent bonds are formed between certain amino acids to stabilize the final structure of the protein molecule. The amino acid sequence of a protein is what determines this structure, as each different R-group in an amino acid has distinct properties. Common proteins, found in most species, will have very similar amino acid sequences and structures. However, a single amino-acid substitution can change the final folded structure of a protein. These changes may be subtle or radical. Closely-related species will share similar amino acid sequences, but as species diverge on an evolutionary scale, more differences in amino acid sequence can be found. With these changes in sequence come changes in protein folding and structure.

ISOLATION OF PROTEINS

When you want to isolate proteins from a cell or tissue sample, you can easily create a crude protein extract by grinding the tissue and then resuspending in an isotonic buffer (like phosphate-buffered saline). After centrifugation of the crude extract, the soluble proteins will stay in solution, while the insoluble proteins will fall out of solution and form a solid pellet. The soluble proteins can be loaded onto a chromatography column or a PAGE-gel, but the insoluble proteins will have to be solubilized before loading. This is typically accomplished by adding a new buffer containing detergent (like SDS, Triton X-100) to the pellet. The detergent will help to free the proteins from membranes by which they may be bound. Some non-polar, hydrophobic proteins may require solubilization with other organic compounds.

SDS—POLYACRILAMIDE ELECTROPHORESIS

Proteins in solution will migrate in an electrical field according to their net charge, size, and shape. Separation of molecules by movement in an electrical field is termed electrophoresis. In SDS-PAGE, proteins are first denatured by treating them with sodium dodecyl sulfate (SDS), a negatively-charged detergent. The hydrophobic portion of SDS interacts with the hydrophobic domains of proteins, causing them to partially unfold or denature. Along with heating, this causes the proteins to all take on the same, linear shape.

An Electrophoresis Unit

After treatment with SDS, a protein is surrounded with a large number of negative charges that overwhelm the protein's intrinsic charge(s). Therefore, all proteins take on the same charge. The negatively charged proteins are subjected to electrophoresis through a highly cross-linked matrix of acryl amide (termed a gel). By giving all sample proteins the same relative shape and charge differential, movement of the proteins through the gel is based primarily on size, as the linear molecule weaves its way through the gel matrix. The electrical field is configured such that the negatively charged proteins migrate through the gel from a negative electrode toward a positive electrode.

Polyacrilamide gels are long, cross-linked polymers produced with varying pore sizes and are widely used to separate mixtures of proteins. The gel pore size is affected by the total acryl amide concentration. This permits the electrophoretic medium to have a direct effect on the separation of molecules of different sizes. If the pore sizes are approximately the same size as the molecules, smaller molecules will move more freely, and the larger molecules will be restricted in their migration.

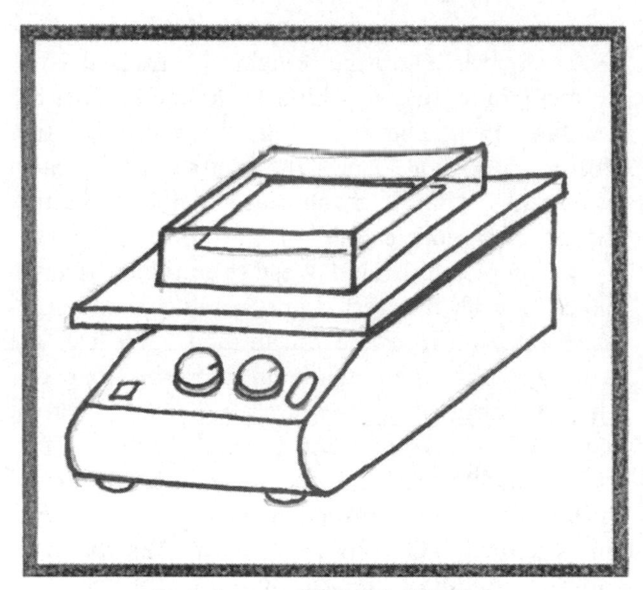

Staining an SDS-PAGE Gel

To visualize the proteins, the gel is stained with a protein-binding dye after electrophoresis is completed. A series of molecular weight (MW) standards will separate in the SDS gel according to the MW of the individual proteins. One can estimate the MW of the proteins run on the same gel by comparison to the position of the MW markers in the gel. The MW marker sizes in kilodaltons are shown in the figure. Because each protein will move through the gel based on size, proteins with specific molecular weights can be mixed together and used for size analysis as one form of control on an SDS-PAGE gel.

Since SDS denatures proteins, a protein run on an SDS-PAGE gel is not functional. However, protein samples can be separated on a PAGE gel lacking SDS. These Native PAGE gels will create a different pattern of bands, since proteins will be separated based on molecular weight and charge. Typically, Native PAGE gels are used to extract sensitive protein samples from the gel itself by cutting a protein band out and purifying it from the polyacrilamide.

WESTERN BLOT & IMMUNODETECTION

Proteins in a gel can be transferred to a nitrocellulose membrane, and a specific protein may be detected using antibodies that bind to that particular protein. The membrane that contains the transferred proteins is termed a Western, or immuno, blot, and the overall procedure is termed Western Immunodetection. Proteins can be transferred to the membrane using capillary action or electrophoresis. Once the proteins are transferred to the membrane, they will bind to the membrane, but maintain their chemical and structural identity. This is important, since the shape of a protein molecule is what is recognized by the antibodies used in immunodetection.

The principle behind Western immunodetection makes use of the very tight binding that occurs between an antigen and an antibody. We will use antibodies that were made in a rabbit that was injected with protein from a crab. When the rabbit had an immune response to the crab protein, it created antibodies specific to that protein. Most of these antibodies are of a particular class of immunoglobulins termed IgG. These IgG antibodies recognize only the protein of interest, and not other proteins, and are thus "specific." These first or primary antibodies are washed onto the membrane, where they can then interact with and tightly bind to the protein band from your crab sample.

The antigen-antibody complex is then identified by a second step of immunological detection. This involves incubating the membrane with a "secondary" antibody that binds to the "primary" antibody (e.g., a mouse anti-rabbit IgG, created by injecting IgG proteins into a mouse). The secondary antibody does not bind to the protein of interest; instead, it binds to the primary antibody. The secondary antibody is typically conjugated to a molecule used to detect the antigen-primary antibody-secondary antibody complex. This molecule may be a radioisotope, which could be detected using X-ray film. It could be a fluorescent chromophore, which could be excited to emit light when placed under an ultraviolet light. In our case, an enzyme is conjugated to the secondary antibody. This enzyme is used in a colorimetric assay. When the blot is washed in the assay buffer, the enzyme conjugated to the secondary antibody will create a colored product. This will help us to identify protein bands specific to our antigen.

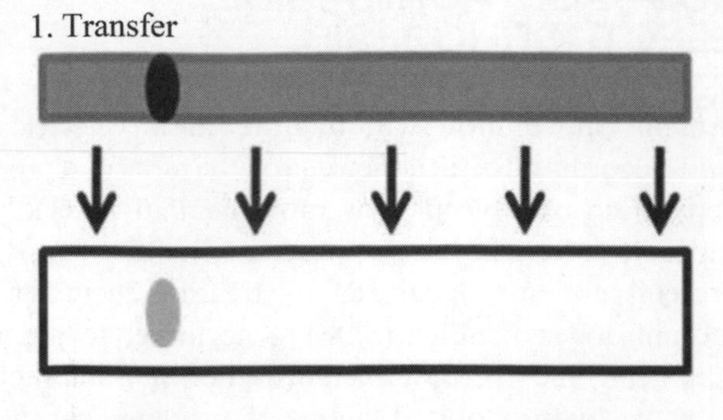

2. Blocking

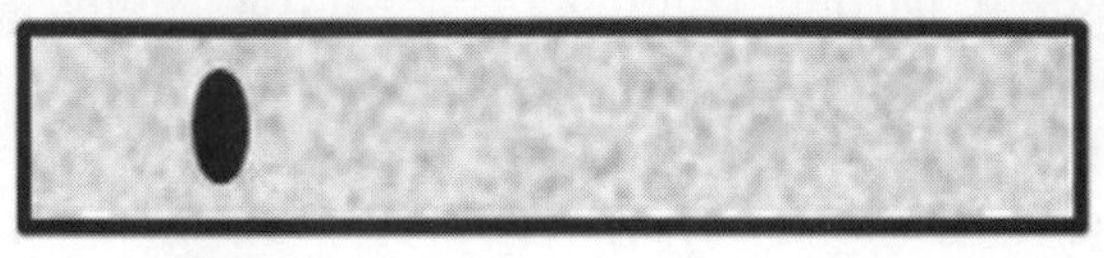

3. Primary antibody

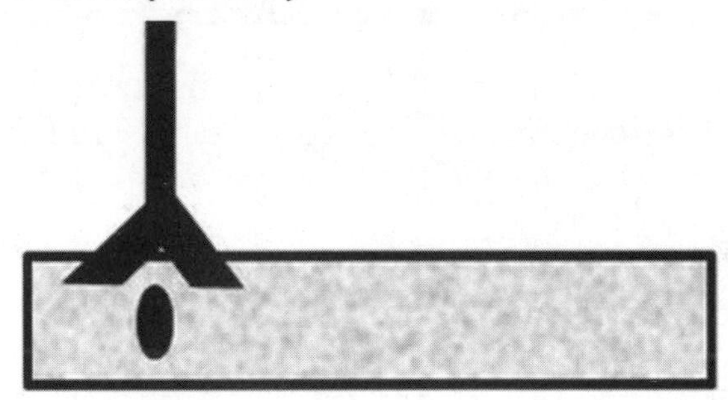

4. Secondary antibody

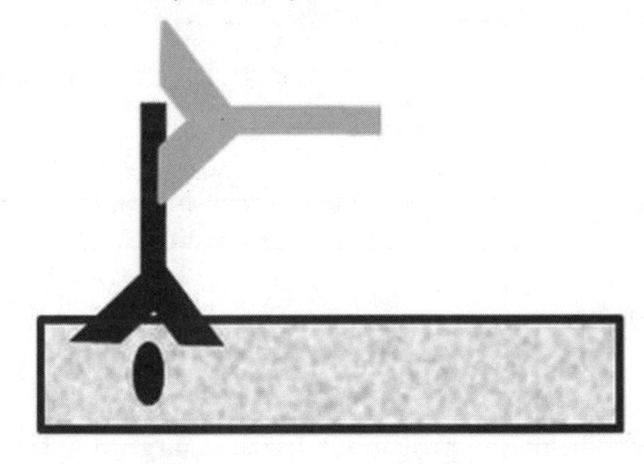

5. Detection

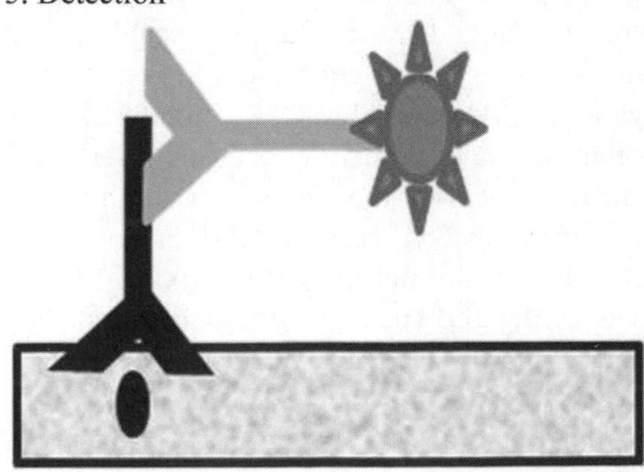

ANALYSIS & COMPARISON

When you run a crude protein extract on an SDS-PAGE gel, you are running almost all of the proteins from the source tissue in a single lane of the gel. After staining the gel, you will see a ladder of bands of varying sizes. Each band should represent a single kind of protein, but if there are any proteins of similar sizes, their bands will blend together. Also, highly concentrated proteins will create darker, larger bands on the gel. In a purified protein sample containing a single protein, you will only see one band on an SDS-PAGE gel.

To compare two protein samples for similarities, you need to look for similar banding patterns. Animals of the same species should have relatively similar patterns of heavy and light protein bands on an SDS-PAGE gel. Closely-related species, like Alaskan red king crabs and Alaskan blue king crabs will have only slightly different band patterns. The further removed the speciation, the more different the pattern. Now, even though most sea life will contain similar kinds of proteins, these proteins will differ in amount, amino acid sequence, molecular weight, and occasionally function. For example, imagine the actin and myosin proteins found in muscle. The types of actin and myosin proteins found in a trout would differ from those found in a clam. This has to do with function, activity, and evolutionary divergence. Even though at a molecular level, the proteins are performing the same basic function, there will be minor differences in more distantly-related species.

If you try to compare vastly different species, such as land animals versus marine animals or animals versus plants, you will begin to find extreme differences in the protein banding patterns. So, when soy proteins are used as bulk protein filler in ground meat, this makes it very easy to detect the additives with a PAGE-gel. The plant proteins create a band pattern that is very different from patterns found in samples made from animal tissues.

Western blotting with immunodetection is even more specific. Rather than rely on the patterns created by differing molecular weights and concentrations of proteins, the antibodies used to probe the Western blot are specific. They are species-specific and molecule-specific. Now, antibodies will bind to similar proteins within the same species. They may also bind to similar proteins in closely-related species. However, the hybridization specificity decreases with different proteins, or homologous proteins from different species. Less primary antibody will bind to the protein bands, as specificity decreases. This can be quantitatively observed by the amount of secondary antibody that is detected in the colorimetric assay used to analyze the Western blot. The more color you observe, the more specific the hybridization.

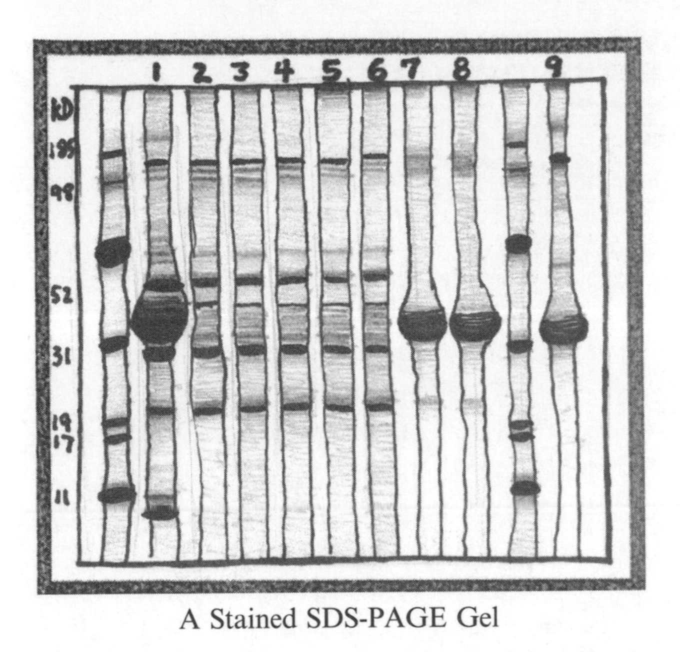

A Stained SDS-PAGE Gel

For example, if you were to grind up protein samples from king crab, shrimp, codfish, and soy, separate them with an SDS-PAGE gel and a Western blot, and use crab-specific antibodies to analyze your blots, what would you observe? When compared with a crabmeat control, you would find that the king crab protein band has the darkest coloration. The shrimp band in the next lane would be lighter than the crab, but darker than the cod band. You would find very little color in the soy band. With evolutionary distance, there is less similarity between homologous proteins and less likelihood of a protein homolog, in general.

PROCEDURES

Protein Isolation and Differential Centrifugation

1. Obtain 1 × 1-cm cube of known tissue. Your instructor should have two different animal protein and one soy protein control available. Choose one for your preparations.
2. Weigh out 1.0 g of fresh tissue, and place it in a mortar and pestle.
3. Add 3.0 mL of cold, deionized water.
4. Grind the mixture.
5. Add 3.0 mL Sample Preparation Buffer. Stir the mixture with a plastic spoon.
6. Scrape the sample into a 15 mL centrifuge tube. Label this tube "insoluble."
7. Rinse the mortar and pestle thoroughly to avoid cross-contamination.
8. Repeat Steps 1 to 6 for your meat sample.
9. Centrifuge 5 minutes to pellet the insoluble solids.
10. Use a transfer pipette to transfer the liquid above the fresh tissue pellet into a new 15 mL centrifuge tube. Label this tube "soluble."
11. Add 3.0 mL of SDS-Buffer to pellet. Vortex and centrifuge 5 minutes.

Sample preparation

12. Add 30 μL of each of your soluble and insoluble sample into a microfuge tube.
13. Add 6 μL of (blue) loading buffer, and incubate them at 70°C for 10 min.
14. Spin the samples in a microcentrifuge briefly to bring the sample to the bottom.

SDS-PAGE Gel Setup

The setup of the gel apparatus and loading will be demonstrated.

1. Add 30 μL of each sample to gels as shown in the figure below.
2. You will share your gel with other students. The instructor will load a pre-stained molecular weight standard.

Run the gels at 150 Volts for approximately 1 hour.

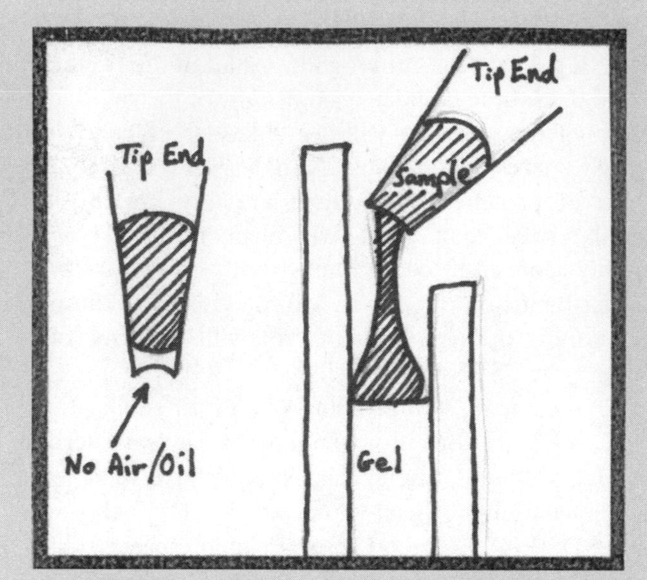

Loading Your Sample onto an SDS-PAGE Gel

Staining and De-staining

1. Add 15 mL of blue stain.
2. Shake on an orbital shaker for 20 minutes.
3. Discard the stain down the sink.
4. Rinse twice with water.
5. Add water to wash on an orbital shaker for 5 minutes.
6. Discard the water down the sink.

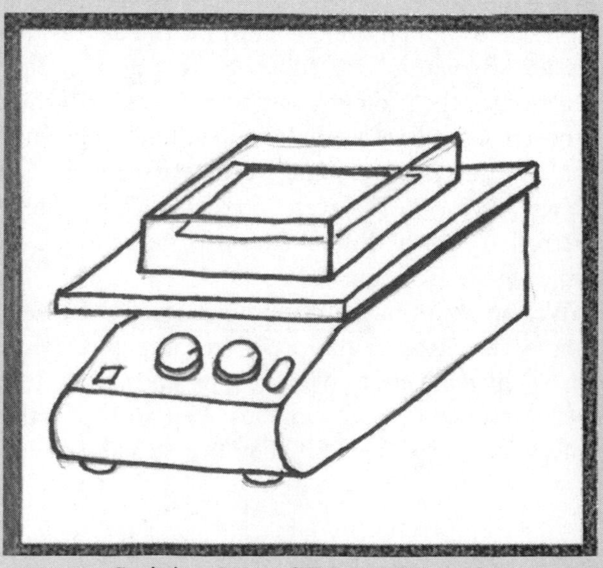
Staining Your SDS-PAGE Gel

17 OVERVIEW

WORKSHEET

Protein Isolation, Fractionation, and SDS-PAGE

1. Define the terms *crude protein extract*, *soluble protein sample*, and *insoluble protein sample*. Explain what kinds of proteins might be found in each. Give at least one specific example for each.

2. Draw a picture (or tape a photograph) of your SDS-PAGE gel results below. Label the lanes and samples for each. Be sure to label any important differences and similarities between the samples.

3. What might your PAGE gel results look like if you did not add any SDS to the sample, gel, or running buffer? What could this "native" PAGE gel be useful for?

4. Use the table below to compare the SDS-PAGE results throughout the class. Detail MW-values for similar and dissimilar bands found in experimental vs. control and at least one other set of protein samples.

Sample	Similar Bands	Unique Bands	Potential Ingredients

5. Using the techniques that you have employed today, create a simple experimental design that describes how you would test the meat and fish products coming into a new Midwestern restaurant. Be sure to explain what you will use for samples; what you will use for controls, which methods you will use, and describe expected results for authentic versus imitation meat products.

DISCUSSION

Discussion Questions

1. At a nearby imaginary school, a high school student nearly died after eating a hamburger. The hamburger was tested for bacterial contamination and was found to be negative for bacteria. No one else became ill after eating hamburgers from the same batch. The student has a severe allergy to soybean products. Soy products contain some of the proteins found in soy beans. Explain how a forensic scientist might analyze the hamburger meat to determine what caused the attack. Design an experiment similar to today's method that would test tofu and soy sauce for the presence of soy proteins. Decide what you would use for a control in each situation. Discuss any possible pitfalls with sample preparation and gel loading.
2. The sample preparation buffer did not contain detergents After centrifugation, what kinds of proteins would be found in the supernatant? What kinds of proteins would be found in the pellet? What did you load onto your gels?
3. A heavily-stained protein band is found to be common among all of the marine protein samples, but is absent from the terrestrial protein samples. You know the molecular weight of the band. Describe how you might discover the identity of the protein.
4. Could protein isolation and an SDS-PAGE gel be used to confirm genetic identity in siblings, or paternity? How about Western blotting and immunodetection? Explain why or why not?
5. What is a zymogram? How might a zymogram be useful when studying enzyme activity? Can you think of a disease where a zymogram might be a useful assay?

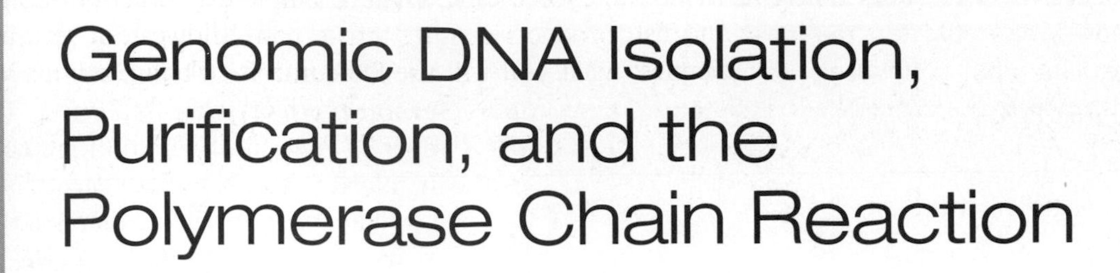

OVERVIEW 18

Genomic DNA Isolation, Purification, and the Polymerase Chain Reaction

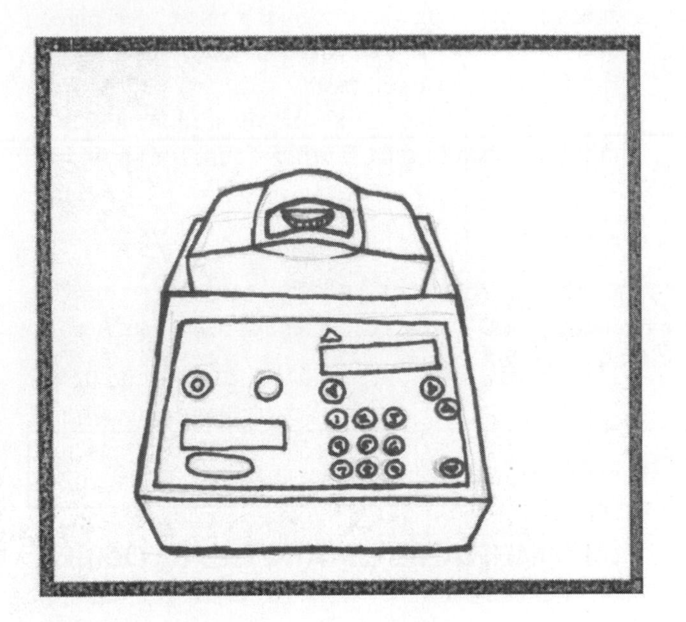

Objectives

- Isolate DNA from cheek cells.
- Amplify a copy of a gene from human genomic DNA.

DESCRIPTION

Recombinant DNA technology provides a method for researchers to isolate specific DNA sequences from a larger, more complex genome. By isolating smaller sequences from the genome, researchers can study gene organization and factors that affect gene expression. The DNA from all of the chromosomes in the cell is called the genomic DNA (gDNA). This gDNA can be extracted and purified by lysing a sample of cells; removing all membranes, proteins and carbohydrates; and purifying and concentrating the remaining DNA. The polymerase chain reaction (PCR) relies on three things: DNA primers specific to the sequence of interest, a thermostable DNA polymerase (Taq polymerase), and thermocycler. From 30 cycles of PCR, a single region of template gDNA can be copied over 1 billion times in a few hours. Today, you will be isolating your own genomic DNA from cheek cells and using PCR to amplify a region of DNA from a specific gene within your genome.

CONCEPTS & VOCABULARY

- Gene
- Genomic DNA
- Lysis
- Melting temperature
- Primers
- Taq polymerase
- Thermocycler

CURRENT APPLICATIONS

- In 2003, the first complete draft of the human genome project was published. The project took over 13 years and $437 million to complete. Today, it takes a month to sequence a genome, and the cost has come down to under $50,000. Within a few years, personal genomics will be readily available to the average American for $1,000 or less.
- PCR has been used to amplify DNA from extinct species, including the dodo, the passenger pigeon and the thylacine (Tasmanian wolf), using laboratory and museum tissue samples. DNA from long extinct animals, like the mastodon, has been amplified using tissue samples found in arctic ice.

REFERENCE

Karp, G. (2010). Purification of nucleic acids and enzymatic amplification of DNA by PCR. *Cell and Molecular Biology: Concepts and Experiments*. (742; 751–753). Hoboken, NJ: John Wiley & Sons, Inc.

BACKGROUND

GENOMIC DNA EXTRACTION

Within every cell of the body, the nucleus holds a complete set of somatic chromosomes. These tightly-bundled structures contain almost the entire cell's deoxyribonucleic acid (DNA), packaged together into coding regions called genes. These genes code for every protein in the body. Some of these genes are only expressed ("turned on") during development, and never activated again. Other genes are constantly expressed throughout the life of an organism. However, these genes can still be found in the nucleus of almost every cell in the body.

Molecular cell biologists often study changes in these genes that affect the proteins that they produce. In order to do this, they must first extract and isolate this genomic DNA from a cell or tissue sample. This DNA can then be used to perform other experiments that will reveal any changes (mutations) in the genes using DNA sequencing. Genomic DNA may even be used to isolate the recombinant genes in order to express their products in an in vitro (cell-free) system or in bacteria. Recombinant proteins can then be used for even further experimentation. By extracting genomic DNA, scientists can match the genotypic changes with the phenotype of the cells or the entire organism, or they can match the change in a gene with the effect on the individual organism.

POLYMERASE CHAIN REACTION

The polymerase chain reaction is a cyclical reaction that amplifies a specific region of DNA. The cycle is composed of three major steps: **denaturation**, **annealing**, and **extension**. In order to amplify a region of DNA, small pieces of DNA are made that will anneal to the single-stranded 5′ and 3′ end of that region. These small DNA molecules are called primers. Primers are added to a template DNA sample, and the mixture is heated to melt or denature the DNA. This is the **denaturation** step. This will convert all DNA molecules from double-stranded DNA (dsDNA) into single-stranded DNA (ssDNA).

Then the mixture is cooled, and the primers will anneal or bind to the 5′ or 3′ end of the region of interest. This is the **annealing** step. This creates a target site for DNA synthesis. Once the primers have annealed, the temperature is raised, and thermostable DNA polymerase (*Taq* polymerase) copies the primed region of DNA through the extension reactions. This is the **extension** step. The mixture is heated again, to denature the DNA samples, and the cycle repeats another 29–39 more times. The thermocycler is a machine that regulates heating and cooling times to allow for fast and efficient PCR. The essential components of this reaction are the template DNA, the primers, the Taq Polymerase, and the thermocycler.

Taq polymerase was the key to PCR. The idea of a cyclical DNA synthesis reaction came earlier, from several researchers, but since normal DNA polymerase would be deactivated with heat, the procedure was too expensive to be effective. However, in 1986, a biochemist named Kary Mullis began using a heat-stable DNA polymerase isolated from an extremophilic bacterium. This addition to the original conception made PCR affordable for every molecular biology lab.

The specificity of PCR comes from the primers. Proper primer design requires a search of available online databases to find specific nucleic acid sequences that will amplify your chosen region of a gene or genome. It also requires calculation of the melting temperature, the heat needed to denature a dsDNA molecule of a specific nucleotide sequence. PCR is semi-conservative. This means that one strand of the original molecule is present in the new dsDNA products. This is because *Taq* polymerase can only synthesize a new strand of DNA when it is attached to an ssDNA molecule, which it uses as a template.

So, with the first reaction you turn one dsDNA molecule into two dsDNA molecules. With the second reaction, two dsDNA molecules become four dsDNA molecules. After 30 reactions, one dsDNA molecule has become over a billion dsDNA molecules. The combination of speed and specificity is what makes PCR one of the key techniques in molecular biology and essential for many of the scientific advances over the past 25 years.

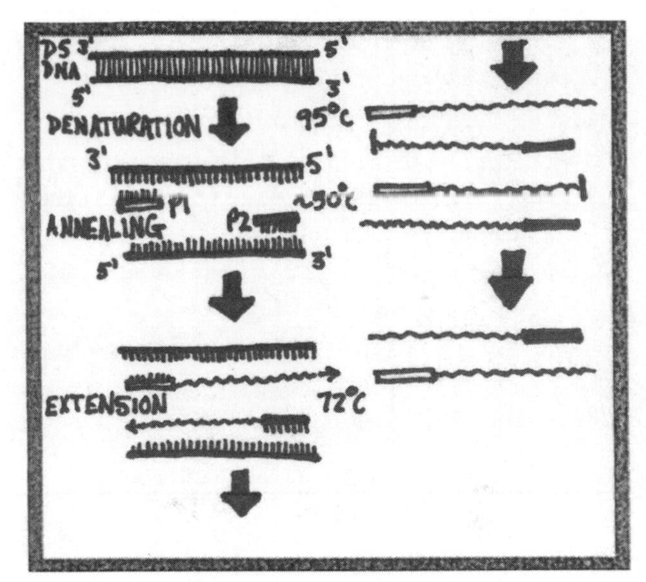

The Polymerase Chain Reaction

PROCEDURES

Genomic DNA Extraction and Isolation

1. Put 600 mL of Lysis Buffer into a 1.5 mL microcentrifuge tube.
2. Gently scrape the inside of your cheek with the flat side of a sterile toothpick.
3. Stir the end of the toothpick in the buffer, and discard the toothpick into the Biohazard Waste.
4. Close lid, and wrap tightly with Parafilm to seal.
5. Boil on a hot-plate or in a water bath for 5 minutes.
6. Spin condensate down briefly in a microcentrifuge.
7. Add 60 mL of Neutralization Buffer and vortex to mix.
8. Centrifuge at 13,000 rpm for 2 minutes.
9. Transfer the supernatant into a fresh tube.

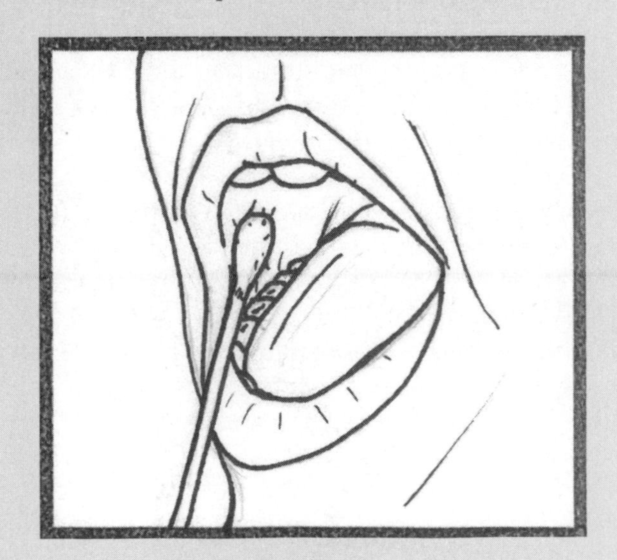

Polymerase Chain Reaction

1. Transfer 3 μL of your gDNA sample to a fresh, thin-walled, PCR tube.
2. Add 22 μL of Master Mix and mix well.
   - The master mix contains your primers, *Taq* Polymerase, magnesium salt, and PCR buffer.
3. Centrifuge briefly to bring sample to bottom of tube.
4. Put your tube in the thermocycler.
   - Before starting the PCR program, your instructors will go through the steps of the program on their machine.
5. Be sure to give your instructor your gDNA sample for storage at 4°C.

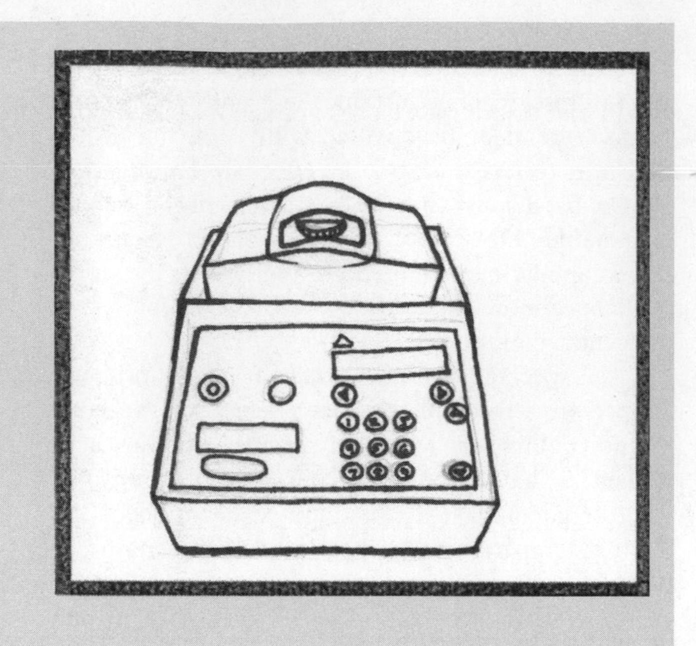

18
OVERVIEW

WORKSHEET

1. What are the steps of genomic DNA isolation? Describe all steps and the importance of each.

2. What other samples could you use for genomic DNA isolation from humans? Which samples would be easiest to retrieve? What samples are available to forensics investigators?

3. Below, draw a diagram that shows one cycle of PCR. Start with one dsDNA molecule and end with two dsDNA molecules. Be sure to label all components and steps.

18 OVERVIEW

4. Students in a different class did not get any PCR products. What suggestions would you make to them for troubleshooting their technique? List the possible things that could have gone wrong and explain how you would prevent this from happening twice.

DISCUSSION

1. Why is it important to lyse the cell? Why is important to neutralize the lysis buffer?
2. List and briefly describe the 3 steps of PCR.
3. You want to study the human salivary amylase gene. Discuss how you would set up a PCR reaction that would amplify this specific gene. Keep in mind that there are two amylase genes in the human body (the other is made in the pancreas).
4. What would happen if you only used a forward primer? Why is *Taq* polymerase necessary?
5. Define the 5′ and the 3′ end of a DNA molecule.

OVERVIEW

19 Ligation and Cloning

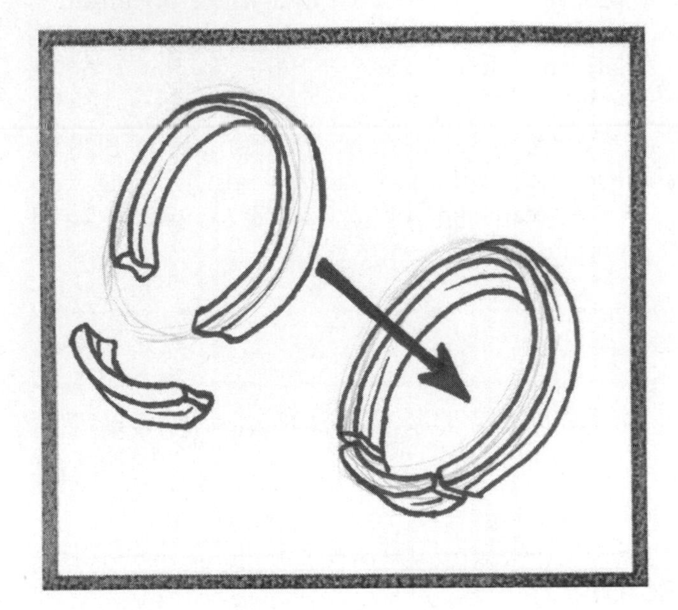

Objectives

- Insert purified DNA fragment into a bacterial cloning vector.
- Insert your ligation into bacterial cells.
- Use antibiotic selection and expression-based screening to identify bacterial colonies that may contain your plasmid and insert.

DESCRIPTION

DNA ligase is an enzyme that can join two pieces of double-stranded DNA together. In the cell, ligase fills in the gaps left by DNA polymerase reactions and DNA repair. In the test tube, ligase is used to "glue" pieces of DNA together. Most commonly, DNA ligase is used to splice a DNA insert into the multiple cloning site of a plasmid vector. Plasmid vectors are tiny, circular self-replicating genetic units similar to eukaryotic chromosomes. While there are plasmids in nature that bacteria use to exchange genes, the plasmids used in the laboratory are artificial. Artificial plasmids are used to replicate large quantities of DNA (cloning vectors) or to direct bacteria to synthesize protein from a foreign gene (expression vectors). Plasmid vectors can be used to change the genetic makeup of bacteria, or transform them. Today, you are going to transform bacterial cells with a cloning vector ligated to your PCR insert. Selection medium will allow you to determine if the cells in each colony contain plasmid and contain plasmid with your DNA insert. Selection puts pressure on the bacteria, which is relieved by some accessory gene on the vector. The most commonly-used selection method is antibiotic-resistance, and that's what we will use today. Antibiotics are added to the medium on the plates, and a matching antibiotic-resistance gene can be found on the plasmid. Those bacterial cells that are successfully transformed with plasmid will grow on the plate, but all other cells will die. Screening methods allow you to determine if a transformed bacterial colony contains your PCR insert. There are a number of screening methods, but the most common involves a blue-to-white color difference. At the next class meeting, you will determine if your ligations and transformations were successful by observing the colonies on your plate.

CONCEPTS & VOCABULARY

- Antibiotic resistance gene
- Cloning vector
- Expression vector
- Ligation
- Multiple cloning site
- *Ori* site
- Plasmid vector
- Screening
- Selection
- T4 DNA ligase
- Transformation

CURRENT APPLICATIONS

- DNA ligation is what makes "gene splicing" possible. Using a combination of restriction digestion to cut DNA and ligation to stick it back together, fusion gene products can be easily created. The most common fusion protein used today is a selection marker called Green Fluorescent Protein (GFP). By fusing a protein that you are interested in to GFP, you can track the location of your protein in a cell or even in a whole organism. If you're interested, today you can even buy GFP-labeled "glow-in-the-dark" fish for your home aquarium.
- Transformation was first demonstrated in 1928 by Frederick Griffith, a bacteriologist working with mice, searching for a vaccine against bacterial pneumonia.

REFERENCE

Karp, G. (2010). Recombinant DNA technology. *Cell and Molecular Biology: Concepts and Experiments.* (748–751). Hoboken, NJ: John Wiley & Sons, Inc.

BACKGROUND

DNA LIGATION AND PLASMID VECTORS

Ligation is a process by which a 5′-phosphate group and a 3′-hydroxyl group on a single strand of DNA are fused together. This reaction is catalyzed in molecular biology by the enzyme T4-DNA Ligase. Using this enzyme, two pieces of DNA can be "glued" together. In molecular cloning, the processes by which bacterial cells are used to replicate large quantities of DNA, plasmid vectors are used. Plasmid vectors are circular, self-replicating DNA constructs that are used to carry foreign DNA into a cell, and may be either cloning or expression vectors. Cloning vectors transform bacteria, which then produce the DNA with every new cell division. Expression vectors transform prokaryotic and eukaryotic cells, which produce the gene product coded by the inserted DNA. There are dozens of commercial plasmids with unique and effective screening and selection protocols. However, for every vector system that you can purchase from a company, researchers have designed a free one. Pay attention to scientific papers, and look for a free vector system that might suit your own research better.

The important parts of a plasmid are the *Ori* site (origin of replication), the multiple cloning site (MCS), and at least one or two selection markers. The *Ori* site is a DNA sequence that allows for the initiation of new DNA synthesis, and therefore allows for replication of the plasmid. The MCS is a grouping of restriction sites, which can be used to join DNA strands together in a specific manner. Remember, if two different pieces of DNA are cut with the same restriction enzyme, they can be joined together specifically at their "sticky ends" with DNA ligase. If a restriction enzyme does not create an overhang, it creates what is known as a "blunt end." Blunt ends can be used for ligations as well, but they lack the specificity of sticky ends. In a bacterial plasmid, successful ligation occurs when both sides of a DNA insert fuse to either end of the linear plasmid vector, thus creating a circularized plasmid containing the DNA insert in the MCS. Your PCR product will be inserted into the MCS of a bacterial vector using T4 DNA Ligase.

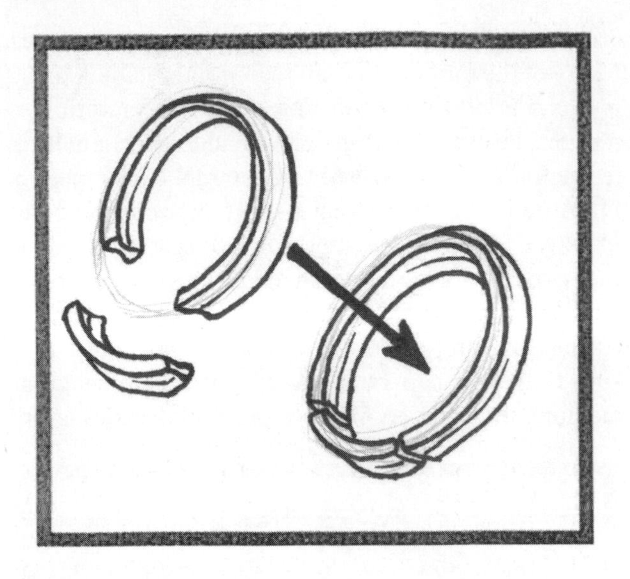

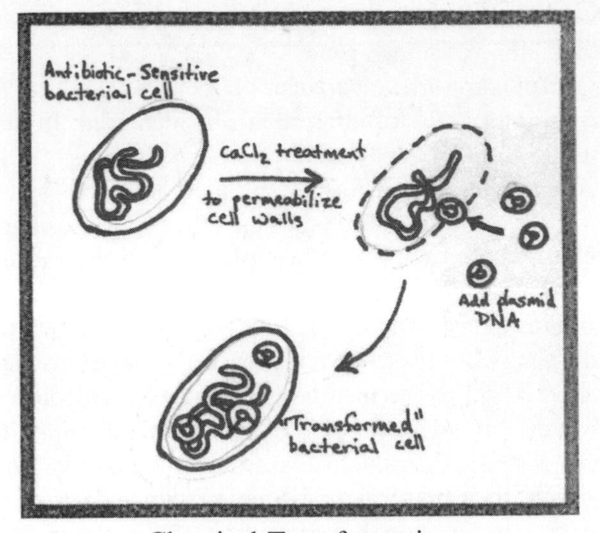

Chemical Transformation

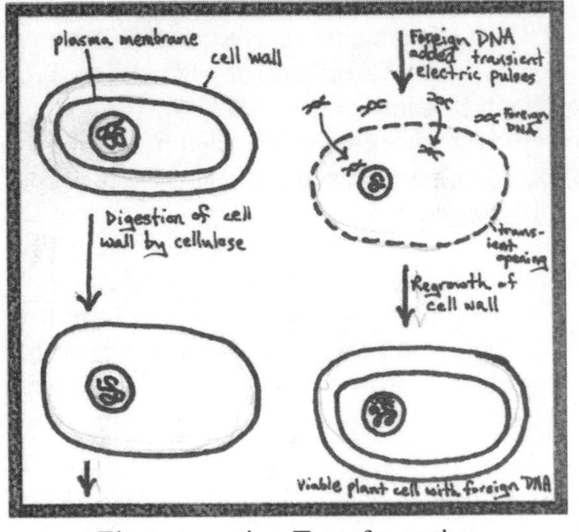

Electroporation Transformation

BACTERIAL TRANSFORMATION

Once you have a ligated plasmid+insert mixture, you are ready to do transformation. Transformation is the process of inserting new DNA into a cell. Transformation relies on uptake of DNA by bacterial cells. This can be achieved by chemical means coupled with a heat shock response, by using electricity to push DNA into the cell, or by using viruses to infect bacterial cells. Eukaryotic transformation can use similar methods, but may also use bacteria to infect cells or may even use gold particles, coated with plasmid, to shoot the DNA into the cell (gene gun). Today, you will be inserting bacterial plasmid into *E. coli* bacteria. There are two forms of bacterial transformation that we will discuss: chemical transformation and electroporation.

Chemical transformation uses salt to alter the plasma membrane of a cell. For example, positively charged calcium ions make the exterior of a bacterium more attractive to negatively charged DNA. The DNA will adhere to the outside of the cell. Then the bacterium can be stimulated to take up new DNA. This is typically accomplished by heat shock, or stress by a momentary incubation at a higher temperature. The heat shock causes DNA uptake, and since the plasmid DNA is adhering to the exterior, it is very likely to enter the cell. After a recovery step in liquid medium, the cells are plated on solid selection medium and allowed to grow overnight.

During electroporation, calcium ions are used to make the bacterial cells more attractive to negatively charged DNA. However, electroporation uses an electric shock to create holes (pores) in bacterial membrane which DNA can travel through. Afterward, during a recovery step, the bacteria will repair these holes and begin to replicate. Again, the cells are plated on solid selection medium and allowed to grow overnight.

SELECTION AND SCREENING

Once a recombinant DNA molecule is formed, it is necessary to transfer the molecule into a living organism in order for it to be replicated. The replication of many identical copies of recombinant DNA molecules by such a host is referred to as "cloning" the DNA. Because most cloning vectors are either bacterial plasmids or bacteriophages, a means to transfer DNA into a bacterial host is required. Your cloned DNA fragment will be identified by the processes of selection and screening.

Selection is any process that puts pressure on a cell that can only be relieved by something artificially inserted in the cell. For example, antibiotic resistance genes are added to the plasmid vector and allow the bacteria to grow on a medium containing that antibiotic. Only the colonies that grow will contain your plasmid. Today we will be using ampicillin and an ampicillin-resistance gene.

Screening is a system of analysis by which the cells containing DNA of interest can be identified. In our case, those are cells that contain our insert. The most commonly-used screening method is the "blue-white" screening method that uses the β-galactosidase gene. A colorimetric method of selection uses a disruption of the enzyme β-galactosidase to determine if your insert is present. If no insert is present, the enzyme complex will break down a modified galactose sugar called X-gal in the medium, producing blue colored colonies. If an insert is present, it will disrupt the production of β-galactosidase, and the colonies will remain their natural white color. Since lactose is needed to stimulate the production of β-galactosidase in this system, but would be broken down by the same enzyme, another modified galactose sugar called IPTG is used to initiate expression. IPTG can cause β-galactosidase expression, but will never be broken down. These selection methods allow you to quickly and efficiently choose the cells that have been transformed. The best screening and selection methods are simple, and inexpensive. Large repositories of DNA inserts can be created from restriction digested genomic DNA fragments inserted into cloning vectors.

Transformation of bacteria allows researchers to create micrograms to grams of DNA or protein, via small- or large-scale fermentation and extractions. Bacterial cells are transformed with a cloning vector or an expression vector, and are used as tiny factories. Many of the modern human recombinant DNA drugs on the market (human insulin or human growth hormone, for example) are grown in bacterial cells in large fermentation vats. While PCR can replicate a piece of DNA billions of times, there is still some cost involved (mostly due to *Taq* polymerase). Bacterial cells will do the same thing at a much cheaper cost (bacterial medium is inexpensive). Plus, bacterial cells can be frozen for long-term storage. Bacteria stored in a −80°C freezer can be streaked onto fresh selection medium and grow to produce plasmid decades later.

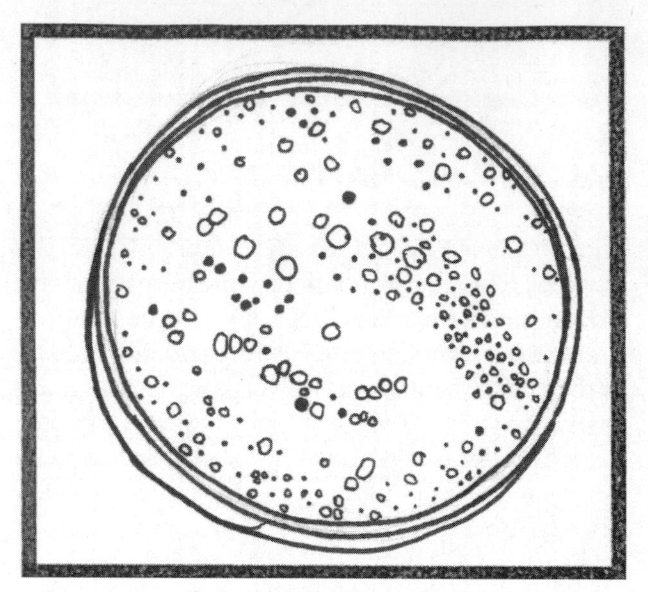

Colorimetric Screening

PROCEDURES

Ligation

1. Obtain your purified PCR fragment from TA.
2. Add 3 μL PCR fragment to the ligation master mix.
3. Incubate at room temperature for 1 hour.

Transformation

1. Take transformation competent cells thaw on ice for 5 minutes.
2. Add 5 μL of ligation mix to 50 μL of cells. Mix gently.
3. Incubate on ice for 30 minutes.
4. Heat shock cells for 90 seconds at 42°C.
5. Incubate on ice for 5 minutes.
6. Add 800 μL of LB Broth and incubate at 37°C for 30 minutes.
7. Plate 100 μL of your cell suspension onto an LB + (Ampicillin, X-gal, IPTG) plate.
   – Your instructor will demonstrate proper plate-spreading technique.
8. Incubate at 37°C overnight.
9. Your instructor will pull the plates and store at 4°C.
10. Discard your cells in the proper waste container.

Selection and Screening

1. Examine your LB + (Ampicillin, X-gal, IPTG) plate. What do you observe?
2. Fill in the data table for question 1 on the worksheet.

WORKSHEET

1. How many blue and white colonies did you get on your x-gal/amp plates? Explain what is happening at the molecular level to produce these results.

   Blue ____________________ White ____________________

2. Describe the main components of a ligation reaction and their importance. What would happen if your cloning vector was still a closed-circular plasmid instead of a cut, linear plasmid? What color would your transformants be if the vector contained the β-gal gene?

3. What would happen to your ligation mixture without the addition of ligase? If you were to run an agarose gel with that ligation mix, what bands would you expect? What if ligase was added? Sketch the gel (don't forget MW markers).

4. What type of results would you expect after transformation if you placed your cells on the following types of agar media: LB only? LB + Ampicillin? LB + X-gal, IPTG?

DISCUSSION

1. Describe the difference between a cloning vector and an expression vector. Which one are you using? Discuss when you would use the other type.
2. Describe the difference between chemical transformation and electroporation. What are the advantages and disadvantages of both?
3. After learning about artificial plasmids and antibiotic resistance genes, discuss how natural antibiotic resistance might proliferate in the natural world. Discuss how antibiotic resistance might proliferate in a hospital setting, where antibiotics are constantly used to treat bacterial infections.
4. Discuss the purpose of X-gal and IPTG when using the blue-white screening system. Why don't we use lactose in the medium?
5. If one of your white colonies on the LB (X-gal, Amp, and IPTG) plate does not contain the desired insert, what would be an explanation for the white color?

Restriction Fragment Length Polymorphism Analysis and Agarose Gel Electrophoresis

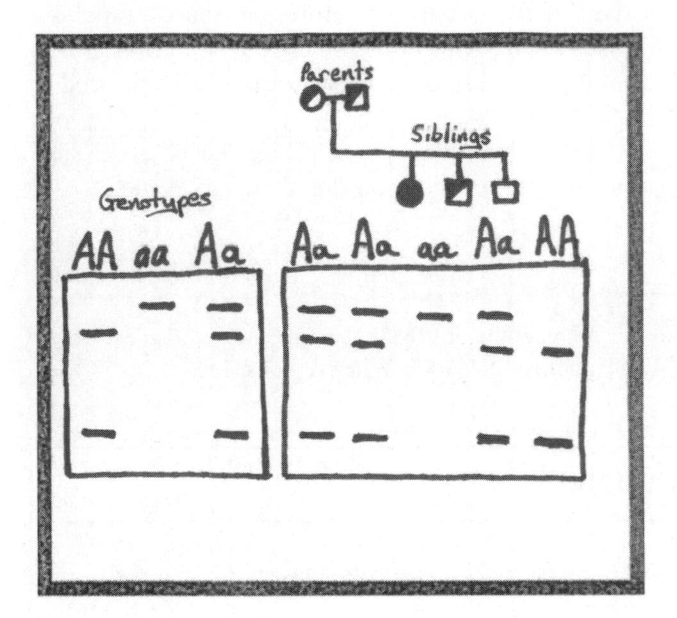

Objectives

- Use restriction enzymes to digest DNA.
- Run the digest on an agarose gel.

DESCRIPTION

Restriction Fragment Length Polymorphism (RFLP) analysis is a technique that can be used to compare the similarities between two DNA samples. RFLP analysis is commonly called DNA testing or DNA "finger-printing." The technique uses a purified DNA sample in which specific sequences of the genomic DNA are amplified by PCR. These specific sequences are well-characterized sequences called amplicons. The amplicons are then cut with a number of restriction enzymes and separated on an agarose gel. The differences in the pattern of bands formed on the gel help researchers to determine the similarity between the samples. In a teaching laboratory, we will only be analyzing one amplicon. This will only give us one RFLP pattern per sample, essentially only the differences in the DNA sequence of a small portion of a single gene. In a comprehensive RFLP analysis, dozens of amplicons from multiple genes are cut with many different restriction enzymes to observe their RFLP patterns.

CONCEPTS & VOCABULARY

- Agarose gel electrophoresis
- Amplicon
- DNA fingerprint
- Ethidium bromide
- Molecular weight marker
- Restriction endonuclease
- RFLP Analysis

CURRENT APPLICATIONS

- RFLP analysis is a cornerstone in forensic applications. By determining the differences in DNA sequence, many levels of identity can be determined. Whether it is individual identity, parental identity, or the identity of long-dead ancestors, DNA samples can be used to determine the genetic relationships of an individual.
- Your mother's mitochondria are passed down to you in the ovum. Therefore, we share mitochondrial DNA with our mother, our maternal grandmother, and so on. Mitochondrial DNA is used to determine ancient maternal ancestry. Men can track their ancestry back down the paternal side, using specific genes on the Y chromosome.

REFERENCE

Karp, G. (2010). DNA fingerprinting, fractionation of nucleic acids and recombinant DNA technology. *Cell and Molecular Biology: Concepts and Experiments.* (395; 742–743; 746–747). Hoboken, NJ: John Wiley & Sons, Inc.

BACKGROUND

RESTRICTION DIGESTION

Restriction enzymes, also known as restriction endonucleases, are enzymes that cut a DNA molecule at a particular site. They are essential tools for recombinant DNA technology. The enzyme binds to a DNA molecule, searching for a particular sequence that is usually of four to six nucleotides long. Once it finds this recognition sequence, it cuts the strands. On double-stranded DNA, the recognition sequence is on both strands, but runs in opposite directions. This allows the enzyme to cut both strands. Sometimes the cut is directly across or blunt; sometimes the cut is uneven with dangling nucleotides on one of the two strands. This uneven cut is known as sticky ends.

BLUNT END

```
5′ NNNNNNNATT          AATNNNNNNNNNN 3′
3′ NNNNNNNTAA          TTANNNNNNNNNN 5′
```

STICKY END

```
5′NNNNNNNNNNNNNNNG             AATTCNNNNNNNNNN 3′
3′NNNNNNNTTAA          CNNNNNNNNNNNNNNNNNNN 5′
```

Most plasmids used for recombinant technology have recognition sequences for a number of restriction enzymes. This allows a scientist to choose from a number of places to cut the plasmid with a restriction enzyme. You always want to read carefully the information sheet that comes with your enzymes as well as the catalog information. The better you know your enzyme, the more likely you will be to have a successful digestion. Most enzymes come in glycerol solution as a storage buffer, but enzymes don't work well in the presence of high glycerol concentration. You want to be sure to dilute the glycerol content down to less than 5% to ensure proper enzymatic activity.

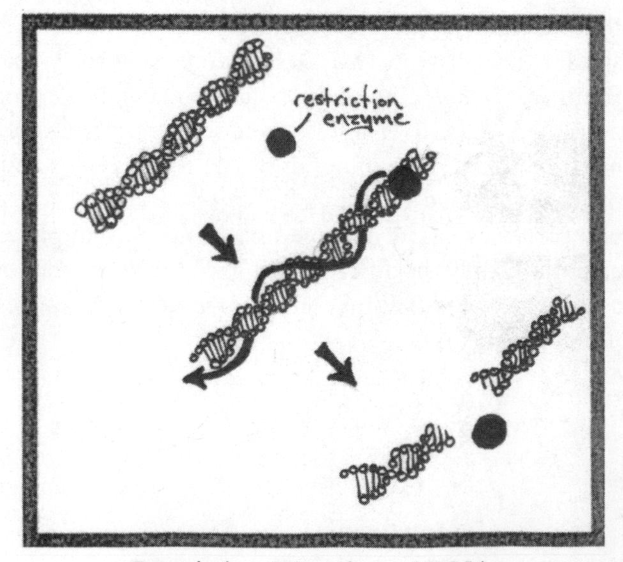

Restriction Digestion of DNA

AGAROSE GEL ELECTROPHORESIS

The most common way to visualize DNA is to separate DNA fragments by size, using agarose gel electrophoresis. In this technique, DNA is separated within a matrix of agarose. An electric current is applied to the agarose (electrophoresis). Since DNA has a net negative charge, it migrates toward the positive electrode. The DNA must weave its way through the matrix. So, smaller fragments move through quickly, while larger fragments are slowed by the agarose matrix. DNA fragments are separated by size with the smallest molecules being closest to the positive electrode. The sizes of the fragments can be determined by comparison with the migration patterns of control DNA fragments of known sizes. Mixtures of DNA fragments of known size, called a molecular weight markers, are often used.

The DNA is then visualized by the stain ethidium bromide (EtBr). EtBr intercalates into the grooves of DNA, and when the gel is visualized on an ultraviolet trans-illuminator, the stained DNA will glow pink. Since EtBr can bind your DNA, it is a mutagen. So be careful when handling it in any concentration.

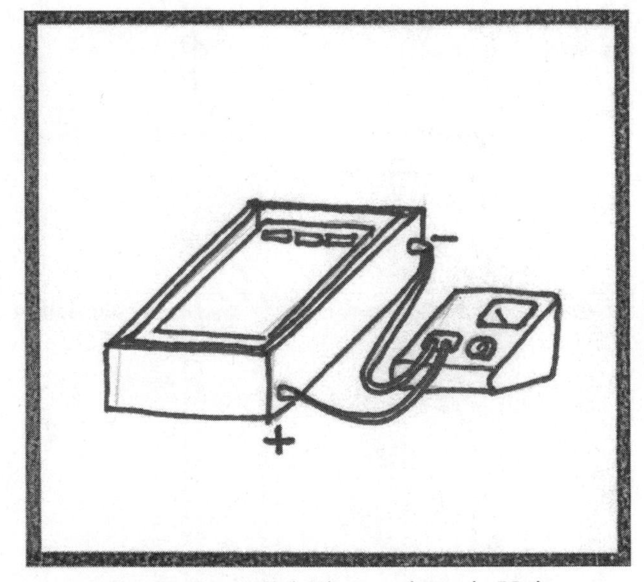

An Agarose Gel Electrophoresis Unit

RESTRICTION FRAGMENT LENGTH POLYMORPHISM ANALYSIS

Restriction fragment length polymorphism is the identification of specific restriction enzymes that reveal a pattern difference between the DNA fragment sizes in individual organisms. To discover RFLPs, restriction enzymes are used to cut DNA at specific 4–6 base pair recognition sites. Sample DNA is cut with one or more restriction enzymes, and resulting fragments are separated according to molecular size using agarose gel electrophoresis. Differences in fragment length result when there are base substitutions, additions, deletions, or sequence rearrangements within the restriction enzyme recognition sequences. If there is enough of a difference to change the recognition sequence, the restriction enzyme can no longer cut at that site. This changes the overall banding pattern.

So therefore, although two individuals of the same species have almost identical genomes, they will always differ at a few nucleotides. Some of these differences will produce new restriction sites (or remove them), and the banding pattern seen will thus be affected. For any given gene, it is often possible to test different restriction enzymes until you find one that gives a pattern difference between two individuals, which is called an RFLP. The less related the individuals, the more divergent their DNA sequences are, and the more likely you are to find a RFLP. RFLP is most suited to studies within a species or among closely related species. Presence and absence of fragments resulting from changes in recognition sites are used for identifying species or populations.

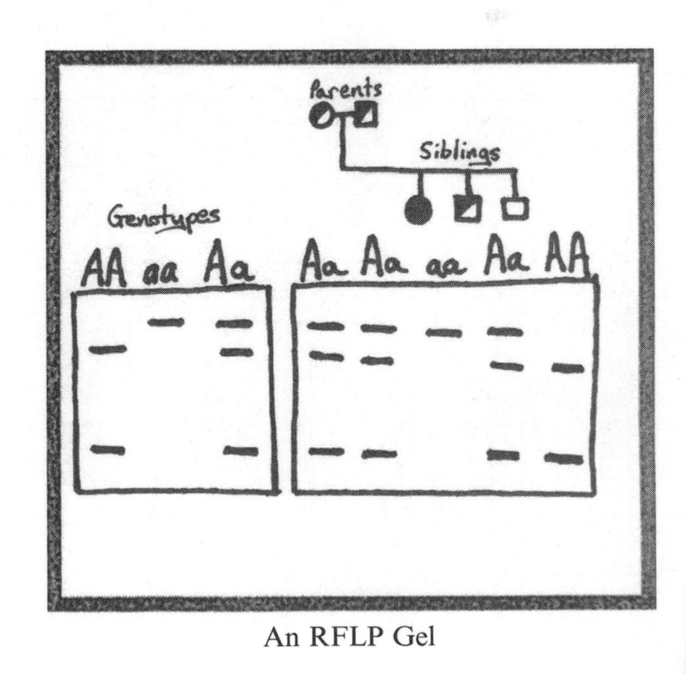

An RFLP Gel

20 OVERVIEW

PROCEDURES

Restriction Digest

1. Label Three 1.5-mL microfuge tubes with C, 1, and 2. Put your name on all tubes.
2. Add 5 μL of PCR product DNA to 5 μL of the enzyme master mix #1 into Tube 1.
3. Add 5 μL of PCR product DNA to 5 μL of the enzyme master mix #2 into Tube 2.
4. Place the tube in a 37°C water bath for 1 hour.
5. Save the remainder of your PCR product DNA on ice.

Agarose Gel Electrophoresis

1. Add 2 μL of tracking dye to 3 tubes (#1, #2, C)
2. Centrifuge for 5 seconds to bring solution to bottom of tube.
3. Submerge your gel into the TAE running buffer.
4. Remove the comb carefully.
5. Working in pairs, load the sample into the wells in the following order:
   - Molecular Weight Marker (Loaded by instructor)
   - Tube "C" – Uncut plasmid
   - Digest #1 ________________
   - Digest #2 ________________
   - Tube "C" – Uncut plasmid
   - Digest #1 ________________
   - Digest #2 ________________
6. Connect the electrodes negative (black) to positive (red).
7. Run the gel at 100 volts for about 1 hour.
   - Be careful of the high voltage.
   - Until dye migrates to about 2 cm from bottom of gel.
8. Remove gel, and give it to your instructor to stain with ethidium bromide.
   - REMEMBER EtBr is a CARCINOGEN.
   - Wear gloves.
   - The EtBr will allow you to visualize your DNA.
9. Visualize and photograph your gel on the trans-illuminator and gel documentation unit.

Loading an Agarose gel

20

OVERVIEW

WORKSHEET

Restriction Fragment Length Polymorphism and Agarose Gel Electrophoresis

1. Attach a copy of your PCR gel picture. What is the molecular weight of your amplified fragment? What are the molecular weights of your first and second digests? Be sure to label all lanes.

2. Look through the class data. Did everyone get the same banding patterns for Digest #1 and Digest #2? Explain the possibilities for the observed similarities, and discuss reasons for any differences.

____

____

____

____

____

____

____

3. Plasmid DNA in bacteria is circular. On an agarose gel, plasmid DNA typically creates two banding patterns because the circularized DNA can coil back on itself. Typically, one band is uncoiled, and one band is supercoiled, with the supercoiled band running slightly faster because it is a more compact molecule. What kind of banding pattern would you expect if you made a single restriction endonuclease cut in the plasmid DNA? Draw your expectation into lane 3 on the figure below.

4. DNA samples were collected from a mother, child, and three possible fathers. Then the DNA samples were analyzed via RFLP. Based on the gel electrophoresis picture below, who is the father and why?

DISCUSSION

1. What characteristic(s) of DNA allow for separation in an agarose gel? Explain the importance of using a molecular weight marker for restriction mapping.
2. In your own words, describe how plasmids and restriction endonucleases are useful for researchers today.
3. Predict what kind of DNA band pattern you would see if RFLP was performed on DNA collected from a set of identical twins. Is there any genetic evidence that could determine a difference between two identical twins?
4. Can RFLP be performed with a strand of mRNA? Why, or why not?
5. RFLP patterns are determined by single nucleotide changes in a DNA molecule. Discuss what causes these changes and where in the genome these changes might be found.

Staining of Peroxisomes

Objectives

- Observe prepared cell samples with stained peroxisomes, using UV-fluorescence microscopy.
- Prepare and observe peroxisome staining in fixed cell samples.

DESCRIPTION

The peroxisome is a specialized, but ubiquitous organelle, required for some of the more advanced metabolism found in eukaryotic cells. Peroxisomes are smaller than lysosomes, and they are filled with dense, dark-colored crystals. These crystals are aggregates of oxidative enzymes. Within the lumen of the peroxisome, a number of metabolic processes occur; for example, the oxidation of fatty acids. This process is separated because fatty acid oxidization creates reactive oxygen species, which could do damage to most other molecules in the cell. The deactivation and degradation of reactive oxygen species also occurs in the peroxisome. There are also enzymes involved in the synthesis of hydrogen peroxide. The peroxisome also contains the enzyme catalase. Catalase is used to break down hydrogen peroxide quickly into oxygen and water. Most people, who want to characterize the peroxisome use immunohistochemical staining. With immunohistochemical staining, they can look for the presence of antigens specific to the peroxisome. One common example of an antigen used in the labeling of peroxisomes is peroxisome membrane protein 70 (PMP-70). PMP-70 is found on the external surface of the peroxisome membrane. Antibodies specific to this protein can be used during a hybridization reaction of a fixed cellular sample. After the hybridization, and possible second hybridization, the cell samples can be observed using UV fluorescence microscopy or confocal microscopy in order to identify the presence and observe the structure of the peroxisome.

CONCEPTS & VOCABULARY

- Antibodies
- Antigen
- Catalase
- Fixation
- Glyoxysome
- Hybridization
- Immunohistochemical staining
- PMP-70

CURRENT APPLICATIONS

- Peroxisomes and their function are essential to the processes that control cell aging and cell death. Proper management of oxidative stress, reactive oxygen species, and free radicals is essential for signaling process and the health of the cell. Since many forms of metabolism produce these harmful, oxidizing molecules, the peroxisome is an essential organelle for eukaryotes.
- Immunohistological labeling is an excellent way to label antigens-of-interest. So long as you are able to fix your sample and purchase or produce antibodies for your chosen antigen, there are dozens of interesting experiments that may be performed. You can identify cellular location of your antigen, determine if it is found in any specific organelles, and find out if it is co-localized with other antigens within the cell. Aside from *in vivo* labeling with fluorescent fusion proteins, this technique is the key to determining the molecular contents of the cell.

REFERENCES

Bonifacino, JS, et al. (2004) Immunofluorescence staining. *Short Protocols in Cell Biology*. Hoboken, NJ: John Wiley & Sons, Inc. (5–22 to 5–24).

Karp, G. (2010). Peroxisomes. *Cell and Molecular Biology: Concepts and Experiments*. (200–203, 225–226, 309). Hoboken, NJ: John Wiley & Sons, Inc.

BACKGROUND

PEROXISOME STRUCTURE AND FUNCTION

Within all eukaryotic cells, there are two populations of small punctate vesicular organelles. The first one is the lysosome; lysosomes are acidic spherical membrane-bound organelles that are filled with digestive enzymes. Similar in size, but different in morphology, are the peroxisomes. Peroxisomes tend to be smaller than lysosomes, and they are filled with dense, dark-colored crystals. These crystals are aggregates of oxidative enzymes. Like the lysosome and the mitochondria, the peroxisome is a separate compartment, and this compartmentalization allows for safety. The oxidative enzymes in the peroxisome would be dangerous to the other parts of the cell. Therefore, the peroxisome is separated by a membrane, and the lumen of the peroxisome has its own unique character. Within the lumen of the peroxisome, a number of metabolic processes occur; for example, the oxidation of fatty acids. This process is separated because fatty acid oxidization creates reactive oxygen species that could do damage to most other molecules in the cell.

On the same note, the deactivation and degradation of reactive oxygen species also occur in the peroxisome. There are also enzymes involved in the synthesis of hydrogen peroxide. Hydrogen peroxide is used by the cell as a strong oxidizing agent. For example, in some of the cells of the liver, hydrogen peroxide is used to modify alcohols, phenols, and formaldehyde, during their conversion to less harmful compounds. Hydrogen peroxide is also used as a signaling molecule, but typically the hydrogen peroxide used in intracellular signaling is produced either in the mitochondria or in a chloroplast in plant cells. Finally and most importantly, the peroxisome contains the enzyme catalase. Catalase is used to break down hydrogen peroxide quickly into oxygen and water. Most bacteria and all eukaryotes use catalase to break down hydrogen peroxide, whether it is being used as a signaling molecule, an oxidizing agent, or as a byproduct of some other form of metabolism. Plant cells also have a unique form of peroxisome called the glyoxysome. Glyoxysomes are used to conduct a specialized form for fatty acid breakdown called the glyoxylate cycle. In the glyoxylate cycle, fatty acids are broken down into acetyl-CoA, then converted into oxaloacetate, then citrate, and then glucose. This allows plants to quickly convert stored fatty acids into carbohydrates.

The peroxisome is a specialized, but ubiquitous organelle, required for some of the more advanced metabolism found in eukaryotic cells.

PEROXISOME STAINING TECHNIQUES

In order to label or identify the peroxisomes, one could use the unique oxidative character of the organelle; unfortunately most of the compounds that could be used to identify a highly oxidative organelle would also be quickly broken down within the peroxisome itself. Also in fixed samples, this oxidative nature is often lost during the fixation process. This is why most people who want to characterize the peroxisome use immunohistochemical staining. With immunohistochemical staining, they can look for the presence of antigens specific to the peroxisome. The majority of antigens chosen are proteins found on peroxisome membranes; however, peroxisomal enzymes and other proteins could be targets for labeling as well. One common example of an antigen used in the labeling of peroxisomes is peroxisome membrane protein 70 (PMP-70). PMP-70 is found on the external surface of the peroxisome membrane. Antibodies specific to this protein can be used during a hybridization reaction of a fixed cellular sample. These antibodies can then be conjugated to a label; usually a fluorescent marker or secondary antibodies specific to the primary antibodies can carry the conjugated label. After the hybridization, and possible second hybridization, the cell samples can be observed using UV fluorescence microscopy or confocal microscopy in order to identify the presence and observe the structure of the peroxisome.

PROCEDURES

Observation of Prepared Peroxisome Stained Sample

While you are waiting for incubations during the next two procedures, take the time to observe a professionally prepared slide that features staining of the peroxisomes. This slide may feature 2–3 other fluorescently labeled cellular elements, so take time to observe all of the available chromophores.

Immunological-Staining of Peroxisomes in Fixed Cell Sample

1. Get two prepared slides from your instructor. Take a moment to determine the top of each slide, the location of the cell sample, and to find the wells drawn on the slide.
2. Your sample is already in blocking solution. Carefully remove the blocking solution with your pipette.
3. Add enough Primary-Antibody solution to cover the sample on one slide. Add a similar amount of PBS/FBS solution to the other slide. Label the second slide "CONTROL."
4. Incubate 1 hour at room temperature.
5. After incubation, add one volume of PBS/FBS, and let sit for 2 minutes. Then remove all of the liquid with your pipette.
6. Repeat this wash twice.
7. Add enough Secondary-Antibody solution to cover the sample on both slides.
8. Incubate 1 hour at room temperature.
9. After incubation, add one volume of PBS/FBS, and let sit for 2 minutes. Then remove all of the liquid with your pipette.
10. Repeat this wash twice.

DAPI-Staining of Nuclei

1. After the last wash, remove the wash buffer, and add enough DAPI-staining solution to fill the well.
2. Incubate for 15 min at room temperature in the dark.
3. Remove the staining solution, and wash the cell sample twice with 1× PBS.
4. Add 100 μL of mounting buffer to the slide, and cover with a cover slip.
5. Examine the samples under the light, and make a note of the appearance on your datasheet. Cells should appear to be clear blobs. Switch over to UV-fluorescence and observe the DAPI and Immunofluorescence staining using the proper filter-cube. Make note of the appearance on your datasheet.

WORKSHEET

1. Observe your pre-stained peroxisome slides using epifluorescence. How do the peroxisomes look? Draw an example sketch in the space provided below. What else is stained in your sample? Using two/three colors, sketch a drawing with all elements in the second box. Remember to label all microscope images with the following: organism, cell type, magnification, and stain; add a scale bar at the bottom right of each sketch.

2. Observe your fresh immunofluorescence-stained slide, and sketch the structures that you see using light microscopy and UV-fluorescence microscopy. Remember to label all microscope images with the following: organism, cell type, magnification, and stain; add a scale bar at the bottom right of each sketch.

3. Observe your fresh unstained CONTROL slide, and sketch the structures that you see using light microscopy and UV-fluorescence microscopy. Remember to label all microscope images with the following: organism, cell type, magnification, and stain; add a scale bar at the bottom right of each sketch.

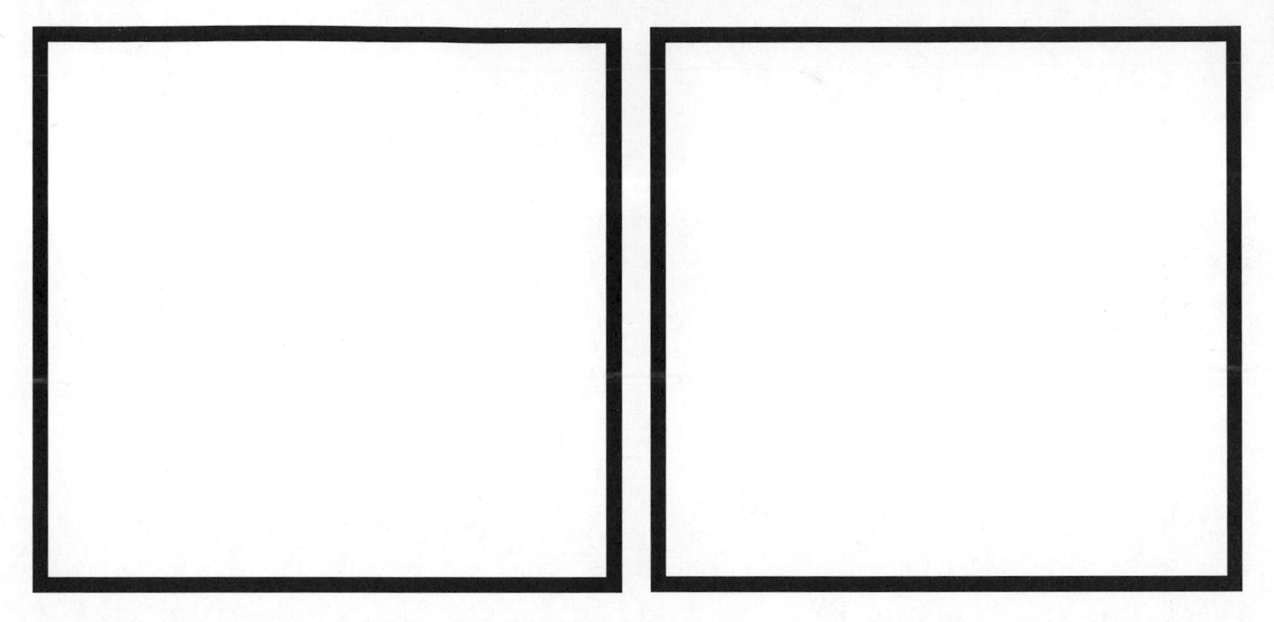

4. How did the professionally prepared slide differ from your fresh preparations? How did your fresh-stained slide differ from your unstained control? Discuss the importance of both of these control samples.

DISCUSSION

1. Discuss the target of antibodies for PMP-70. Did you use a labeled-secondary antibody? How does each antibody assist in the visualization of the peroxisomes?
2. In order to use immunofluorescence labeling, your cell samples must first be fixed. Discuss the reasons for fixation. What would happen if the cells remained unfixed and alive during the hybridization?
3. Plants have two populations of peroxisomes: peroxisomes and glyoxysomes. There are specific antibodies to glyoxysomes, such as antibodies specific to the membrane-bound glyoxysomal lipase. Using this information, discuss an experimental technique that may be used to distinguish plain peroxisomes and glyoxysomes in a fixed plant cell sample.
4. From what you have learned about the function of peroxisomes, what would you expect from a cell line or tissue that produces less or no peroxisomes? What processes would the cell lose in the absence of this organelle?
5. Using the Internet, search for peroxisome disorders in humans. Choose one and discuss the symptoms and characteristics of the disease.

Short-Term Aggregation Culture

OVERVIEW 22

Objectives

- Prepare and observe mammalian suspension cells.
- Prepare and observe mammalian cells in a monolayer.
- Prepare and observe an aggregate culture of mammalian cells.

DESCRIPTION

Cell cultures allow researchers to study the physiology of a specific cell or tissue type outside of the organism. Cell culture has been used for over a century and is currently used for everything from pharmaceutical testing to gene expression analysis. Mammalian cells require a much richer media than would be used for microorganism studies such as bacteria or yeast. Due to the richness of mammalian cell culture media, it is prone to contamination. Specialized hoods are required for mammalian cell culture work, and a more advanced form of the aseptic technique is necessary. However, if you can become adept at the technique of employing cell culture effectively, you can use it as an organism-free (*in vitro*) system. Most modern research laboratories use one of two types of cell culture: primary or secondary cell cultures. Secondary cell culture involves specially prepared frozen cells. Primary cultures are harvested directly from the animal. Cell cultures may be free-floating suspension cell cultures, monolayer, or a two-dimensional cell culture stuck to the base of a culture vial, or aggregate cell cultures that allow for a three-dimensional culture system and a greater variety of cell-to-cell connections. Today in lab, you are going to create an aggregate cell culture and observe the different cell types present in each layer.

CONCEPTS & VOCABULARY

- Aggregate culture
- Cell-to-cell communication
- Cell-to-substrate interactions
- Monolayer culture
- Primary culture
- Secondary culture
- Serum
- Suspension cell culture

CURRENT APPLICATIONS

- Cell cultures are often the first step in human drug testing. After animal testing has been performed, a compound may be introduced to a variety of cell cultures to measure the toxicity and efficacy of the compound. Sometimes, specific cell cultures that display some of the characteristics of diseased cells or tissue can be used to direct test the effects of a drug.
- Cell cultures are also essential for working with viruses. Since viruses are obligate pathogens, they cannot replicate outside of a host cell. Cell cultures allow researchers to maintain a constant pool of virus to work with, without resorting to infecting animals or drawing virus from a human patient.

REFERENCES

Bonifacino, JS, et al. (2004). Short-term aggregation culture. *Short Protocols in Cell Biology.* (13.10–13.12). Hoboken, NJ: John Wiley & Sons, Inc.

Karp, G. (2010). Cell cultures. *Cell and Molecular Biology: Concepts and Experiments.* (743–745). Hoboken, NJ: John Wiley & Sons, Inc.

BACKGROUND

MAMMALIAN CELL CULTURE

Cell cultures allow researchers to study the physiology of a specific cell or tissue type outside of the organism. Cell culture has been used for over a century and is currently used for everything from pharmaceutical testing to gene expression analysis. There are several factors before the cells are even involved that need to be considered before attempting cell culture work.

Mammalian cells require a much richer media than would be used for microorganism studies such as bacteria or yeast. This media is often made from serum from another organism. Most mammalian cell culture serum is produced using serum and proteins extracted from cows. Due to the prevalence of the cattle industry in the United States, bovine protein and serum is readily available. Serum is extracted, cells and cell debris are removed, and then the serum is purified using a complicated filtration process to remove any possible infectious organisms from the serum.

Due to the richness of mammalian cell culture media, it is prone to contamination. Specialized hoods are required for mammalian cell culture work, and a more advanced form of the aseptic technique is necessary to ensure that no foreign organisms are introduced into the culture or the culture hood. The hood itself is often equipped with germicidal lights that use ultraviolet radiation to destroy anything living in the hood. However, if you can become adept at the technique of employing cell culture effectively you can use it as an organism-free (*in vitro*) system.

TYPES OF CELL CULTURE

Most modern research laboratories use one of two types of cell culture. Secondary cell culture is the most common. Secondary cell culture involves specially prepared frozen cell samples that can be stored in a −80°C freezer, or a −120°C liquid nitrogen storage container, for an extended period of time. These frozen cultures can then be revitalized, and the cells can be introduced in a medium and used for scientific research. When cell cultures are shared between laboratories, this is typically the form in which they are delivered.

Primary cultures are harvested directly from the animal. The organ or tissue of interest is removed from the animal via dissection, and then the cells are treated with various protease enzymes, like trypsin. The trypsin will break down the extracellular connective proteins in the tissue and release the individual cells into the media. The extraneous material will be filtered away, and then the remaining cells will be introduced into fresh media. Primary cell cultures can be maintained for a long period of time, so long as they are subcultured before the cells' growth exhausts the medium.

The advantages of one type of cell culture over the other include the fact that secondary cultures are usually tested cultures with known activities, which make studies easy to replicate. Primary cultures allow for greater specificity of cell type choice, but introduce the possibility of randomness or variance into the experiment (depending on how the animals have been cared for). Differences in cage, feed, and even available water may effect the physiology of the harvested cells. In these cases, primary cultures may be converted into secondary cultures to allow for replicate experiments to be performed by other researchers.

CELL ADHESION

Depending on the container, mammalian cell culture forms may or may not bind to the substrate material. If a suspension cell culture is desired, with the cells floating freely in the media, then a non-adhesive substrate is suggested. Suspension cell cultures allow for testing of a specific cell type in a one-dimensional fashion, with each cell having a similar reaction and very little connection or communication with other cells in the culture.

A monolayer or a two-dimensional cell culture may be created by coating the base of a culture vial. A specific coating may be placed on the bottom of the culture vessel to allow for the creation of a single layer of cells. These cells will have two-dimensional connections with each other, which will allow for some cell-to-cell communication, allowing for a more accurate analysis of the physiology of the cells within the tissue. This sort of adhesion is referred to as cell-to-substrate and cell-to-cell adhesion.

Finally, by changing the composition of the media, an aggregate cell culture can be created. This allows for a three-dimensional culture system and a greater variety of cell-to-cell connections. The aggregate culture more closely resembles the structure and positioning of cells within the tissue. Aggregate cultures are bound to the substrate as a monolayer, and then successive layers are bound to each layer of the aggregate creating a multi-tiered culture. As the cells aggregate together, they begin to form the various connections typically found in tissue allowing for communication between cells and the basic set up of the various tissues. Today in lab, you are going to create an aggregate cell culture and observe the different cell types present in each layer.

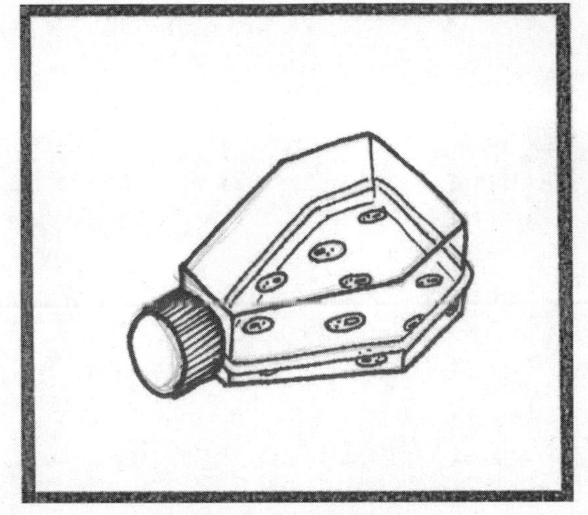

Suspension Cell Culture

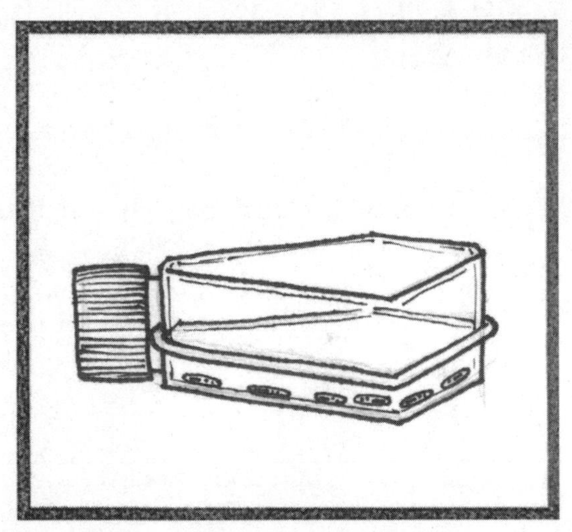

Adherent/Monolayer Cell Culture

Aggregate Cell Culture

PROCEDURES

Initiating Mammalian Cell Culture Aggregation

1. Obtain a vial mammalian cell culture from your instructor. Take it to the culture hood.
2. Examine the sample under the microscope. Make sure that the cells look healthy and intact.
3. Obtain a pre-treated 24-well culture plate from your instructor. A few groups may share a single plate.
4. Label six wells on the plate with your initials and the following labels: A, B, C, D, E, F.
5. Rinse the six wells with 1 mL of HCMF solution three times each.
6. Add 0.5 mL of your cell suspension to each well.
7. Add 56 μL of sterile ddH_2O to wells A, B, and C.
8. Add 56 μL of 100 mM $CaCl_2$ to wells D, E, and F.
9. Take a 1 mL sample from A and D, and place them on ice. Use these samples in the staining procedure below.
10. Seal the plate with micropore tape, and place it on a gyrating shaker at 37°C.
11. After 30 minutes of incubation, take a 1 mL sample from B and E, and place them on ice. Use these samples in the staining procedure below.
12. After 120 minutes of incubation, take a 1 mL sample from C and F, and place them on ice. Use these samples in the staining procedure below.

Staining of Mammalian Cells with DAPI

1. Using a sterile pipette and a wide-bore tip, transfer 1 mL of each cell culture suspension to a separate 15 mL conical tube.
2. Pellet the cells in a clinical centrifuge on setting 2 for 10 minutes.
3. Remove the supernatant, and resuspend the pellet in 5 mL of DAPI staining buffer.
4. Incubate for 15 minutes at room temperature in the dark.
5. Pellet the stained cells in a clinical centrifuge on setting 2 for 10 minutes.
6. Remove the supernatant, and resuspend the pellet in 5 mL 1X PBS.
7. Add one drop of your cell suspension to a microscope slide.
8. Place a cover slip on the slide.
9. Examine the sample under the light microscope with the low-power objective, and make a note of the appearance on your datasheet. Cells should appear to be clear blobs. Switch over to DAPI-fluorescence, and observe the staining. Make note of the appearance on your datasheet.
10. Once you have located a stained, fluorescent cell for each sample, switch to a high-power objective, and examine the cell again. Answer questions 1–3 on your datasheet.

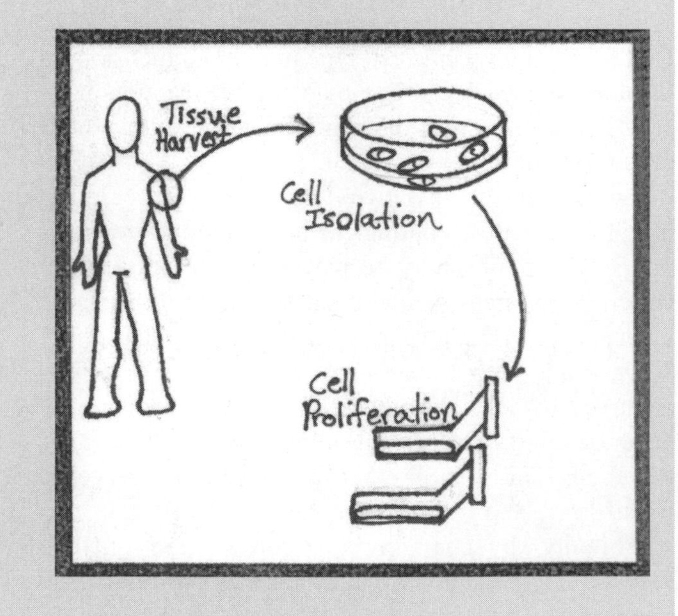

WORKSHEET

Observations and Assessments

1. What does the cell suspension look like after 1 minute of calcium treatment? Draw a sketch of Samples A and D below. Describe any visible structures, and draw an example sketch, using low and high magnification, in the space provided below. Remember to label all microscope images with the following: organism, cell type, magnification, and stain; add a scale bar at the bottom right of each sketch.

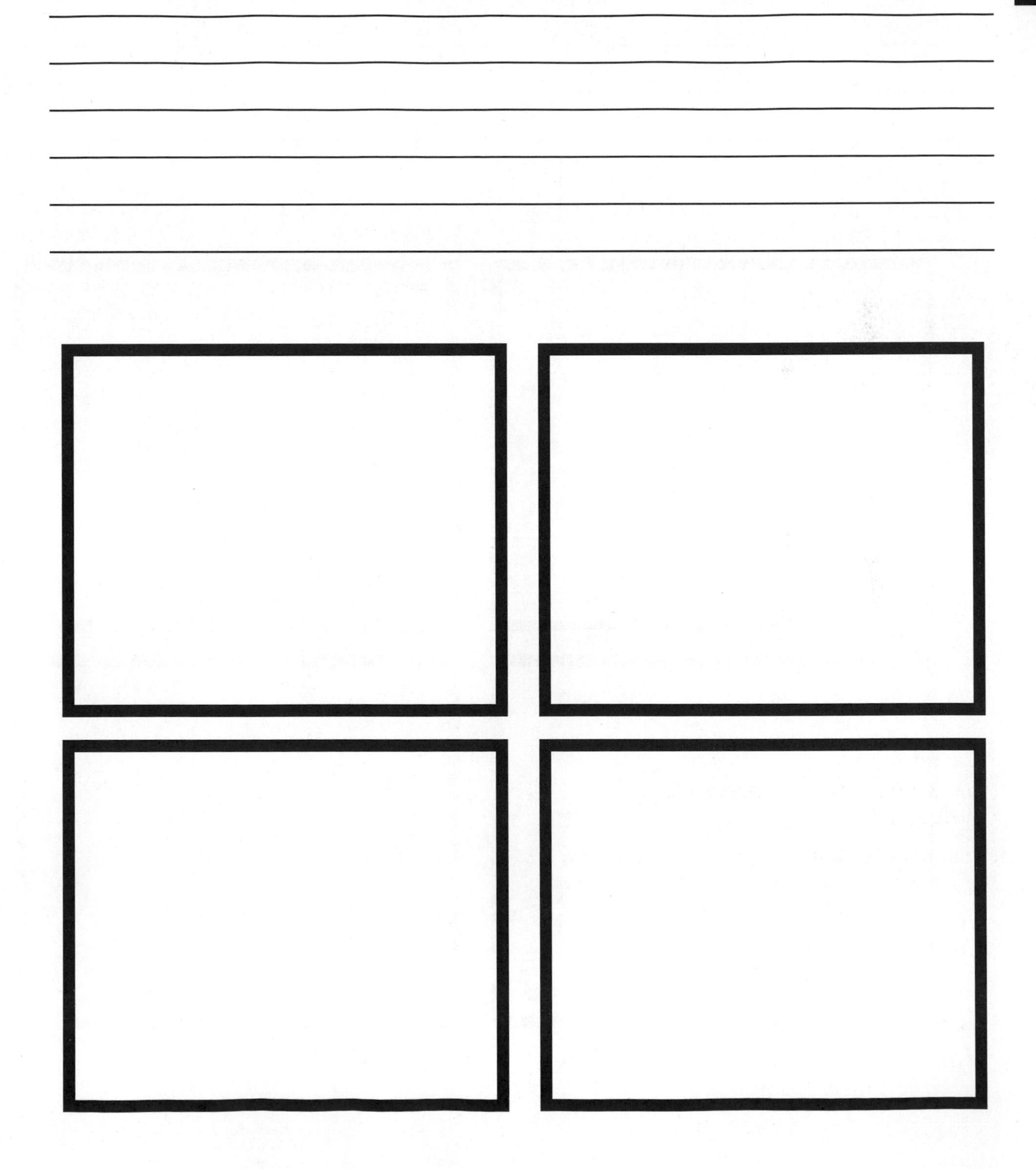

2. What does the cell suspension look like after 30 minutes of calcium treatment? Draw a sketch of Samples B and E below. Describe any visible structures, and draw an example sketch, using low and high magnification, in the space provided below. Remember to label all microscope images with the following: organism, cell type, magnification, and stain; add a scale bar at the bottom right of each sketch.

3. What does the cell suspension look like after 120 minutes of calcium treatment? Draw a sketch of Samples C and F below. Describe any visible structures and draw an example sketch, using low and high magnification, in the space provided below. Remember to label all microscope images with the following: organism, cell type, magnification, and stain; add a scale bar at the bottom right of each sketch.

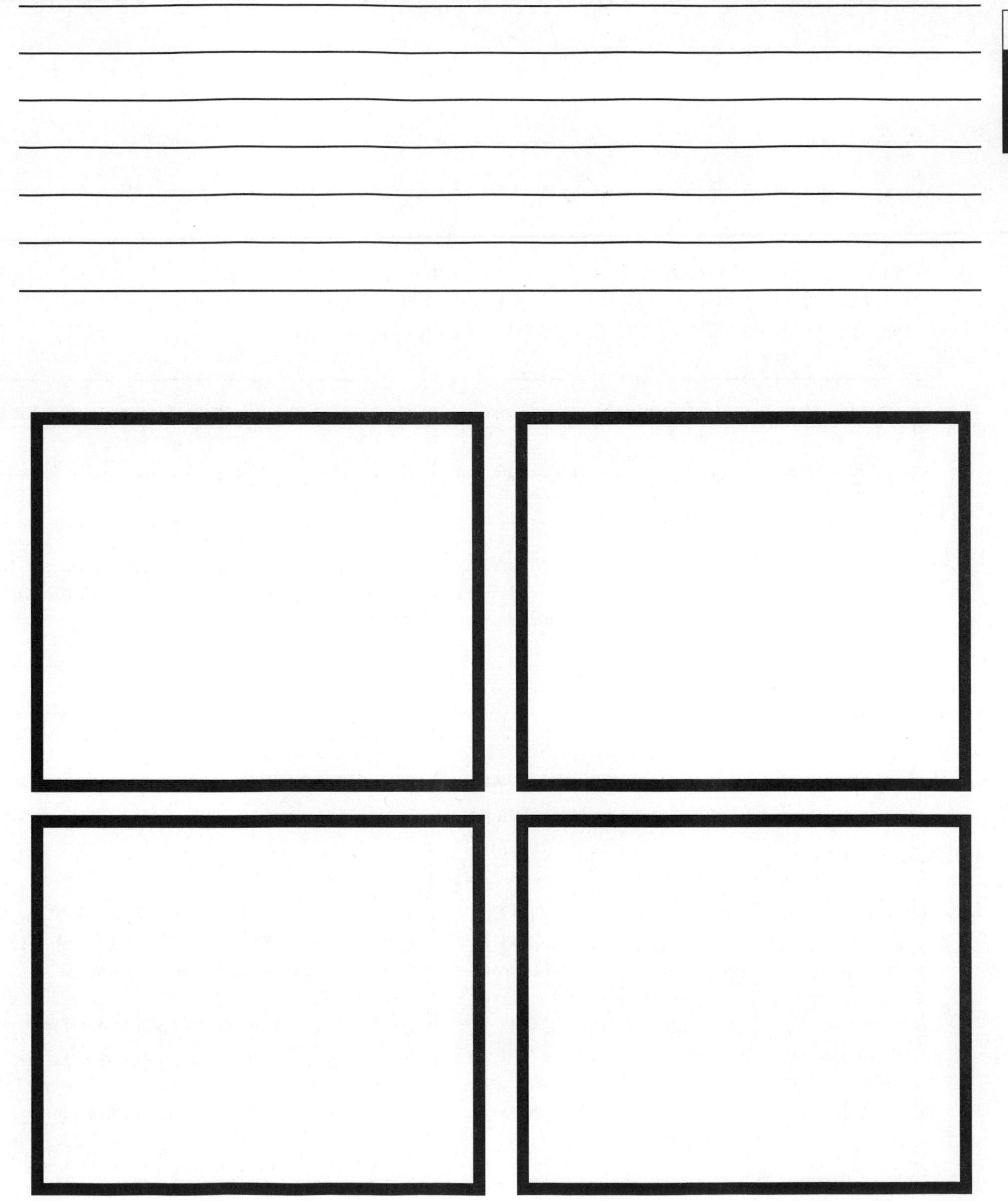

4. Today you used cell culture from trypsinized tissue. What is trypsin? How does it convert tissue into suspension cell culture? How is this process similar to a plant protoplast preparation?

5. What are the benefits of each of the varieties of cell culture (suspension, monolayer, aggregate)? What are the shortcomings of each variety? Which variety would be easiest for the reproduction of results? Which variety would yield results closest to responses found in tissue?

22 OVERVIEW

DISCUSSION

Discussion Questions

1. Discuss possible experimental designs that would only require a suspension cell culture. Discuss possible experimental designs that would only require a monolayer cell culture. Discuss possible experimental designs that would only require an aggregate cell culture.
2. Define cadherins and catenins. How are these molecules involved in cell-to-cell connections and aggregation?
3. The protocol that you used today is only effective with primary cell cultures that have been trypsinized. Discuss why this method might not be effective in secondary cell cultures, or in cultures that have been created using a different enzyme.
4. Calcium is an essential component of aggregation. Discuss the role that calcium plays in culture aggregation and in cell-to-cell connections.
5. DNase I, an enzyme that degrades DNA, has been added to your aggregation medium. DNA from damaged cells tends to inhibit aggregation in culture. Discuss reasons why DNA might prevent aggregation in culture and in tissue.

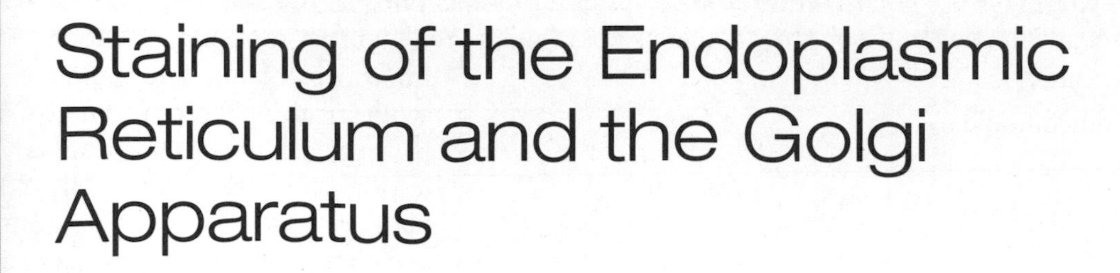

OVERVIEW 23

Staining of the Endoplasmic Reticulum and the Golgi Apparatus

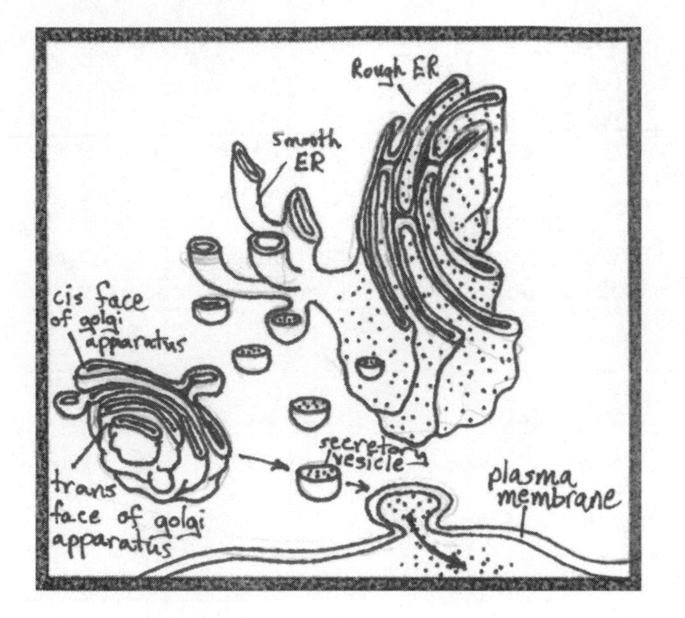

Objectives

- Prepare and observe stained slides of the Endoplasmic Reticulum and Golgi Apparatus.
- Identify the key structures of both organelles.

DESCRIPTION

The endomembrane system of the eukaryotic cell consists of two major parts: the endoplasmic reticulum, which is broken down into smooth and rough endoplasmic reticulum, and the Golgi apparatus, which is divided into *cis*, *medial*, and *trans* faces. Most proteins, membranes, lipids, and carbohydrates are produced at least partly in one of these two compartments. The endomembrane represents the manufacturing unit of the cell. The rough endoplasmic reticulum contains ribosomes. Ribosomes are large riboprotein complexes responsible for the translation of messenger RNA transcripts into amino acid polymers called peptides. Proteins may be also produced within the rough endoplasmic reticulum or within free ribosomes floating in the cytoplasm. However, the nature and function of these proteins depends on their site of synthesis. The structure of the smooth endoplasmic reticulum is more curved and tubular, like pipelines moving through the cytoplasm. The smooth endoplasmic reticulum is the site of synthesis of lipids and steroid hormones. The Golgi apparatus is a modification and packaging organelle. From transcription to final processing, the path is as follows: transcription and processing of a messenger RNA in the nucleus, transport of the message to a ribosome on the rough endoplasmic reticulum or a free ribosome in the cytosol, modification and packaging within the rough endoplasmic reticulum and packaging into a transport vesicle, transport through the cis-, medial- and trans-faces of the Golgi apparatus, final packaging into transport vesicles at the trans-Golgi network. Today you will be observing the morphology of the endoplasmic reticulum and Golgi apparatus using histological procedures specific for labeling of each organelle.

23

OVERVIEW

CONCEPTS & VOCABULARY

- Cisternae
- Endomembrane system
- Ribosome
- Secretory vesicle
- Translation
- Transport vesicle
- Vesicle budding
- Vesicle targeting and fusion

CURRENT APPLICATIONS

- Improper folding and modification of peptides synthesized by the ribosomes of the rough endoplasmic reticulum can lead to several chronic health conditions. A single misfolded protein in the body can cause disease, but ubiquitous protein misfolding can fill the ER with inactive proteins leading to a swelling of the organelle and a condition called ER-stress. ER-stress will eventually lead to cell death.
- The understanding of enzyme and membrane systems employed in the maturation and transport of peptides via the Endomembrane system is a subject of active research. Beyond the identification of the enzymes involved in modifications of peptides, the packaging of peptides into membranes or vesicles, the targeting and transport of those vesicles throughout the cell are not well understood.

REFERENCE

Karp, G. (2010). The endomembrane system. *Cell and Molecular Biology: Concepts and Experiments.* (274–306). Hoboken, NJ: John Wiley & Sons, Inc.

BACKGROUND

STAINING THE ENDOPLASMIC RETICULUM

THE ENDOPLASMIC RETICULUM

The endomembrane system of the eukaryotic cell consists of two major parts: the endoplasmic reticulum, which is broken down into smooth and rough endoplasmic reticulum, and the Golgi apparatus, which is divided into cis, medial, and trans faces. Most proteins, membranes, lipids, and carbohydrates are produced at least partly in one of these two compartments. The endomembrane represents the manufacturing unit of the cell.

The endoplasmic reticulum can be separated into the rough ER and the smooth ER. These designations were given because when the endoplasmic reticulum is separated in a homogenated cell sample, the vesicles of the smooth endoplasmic reticulum appear to have smooth membrane faces, while the vesicles of the rough endoplasmic reticulum have fuzzy or rough membrane faces. This is because the rough endoplasmic reticulum contains ribosomes. Ribosomes are large riboprotein complexes responsible for the translation of messenger RNA transcripts into amino acid polymers called peptides. The rough endoplasmic reticulum is pockmarked with millions of ribosomes, each attached to the cytosolic side and the luminal side of the rough endoplasmic reticulum membrane.

The morphology of the rough endoplasmic reticulum is a complex of flattened sacs or cisternae, which are membranous sacs that contain an interior space or lumen. The luminal or cisternal space has a different molecular character from the cytoplasm or cytosolic space on the outside. The membranes of the rough endoplasmic reticulum are contiguous with the nuclear envelope, and within the rough endoplasmic reticulum protein, carbohydrate and membrane synthesis occurs.

In the smooth endoplasmic reticulum, the structure is more curved and tubular, like pipelines moving through the cytoplasm. The smooth endoplasmic reticulum is the site of synthesis of steroid hormones. In the liver, it is also a point of detoxification. There are many enzymes responsible for the removal of toxins found in the smooth endoplasmic reticulum of cells within the liver. Within muscle cells, a specialized form of the smooth endoplasmic reticulum called the sarcoplasmic reticulum is responsible for calcium segregation and storage. When muscle fibers are excited, this calcium is released, causing the beginning of the energetic cascade that leads to the movement of the entire muscle fiber.

Proteins may be also produced within the rough endoplasmic reticulum or within free ribosomes floating in the cytoplasm. However, the nature and function of these proteins depends on their site of synthesis. Proteins synthesized in the rough endoplasmic reticulum are mainly secreted proteins, integral membrane proteins, or soluble endomembrane proteins found in the luminal or cisternal space. Proteins synthesized on free ribosomes may be cytosolic proteins, or they may be peripheral membrane proteins that are found only on the cytosolic side of the plasma membrane; and proteins found in the nucleus, peroxisome, chloroplast, and mitochondria are also synthesized in free ribosomes.

Proper delivery of synthesized proteins typically utilizes a signal sequence. Signal sequences are a short sequence of amino acids usually found at the amino terminus of a peptide chain that either directs the insertion of a peptide into a membrane or into the luminal space of a vesicle. Further processing performed by the rough endoplasmic reticulum includes the glycosylation of asparagine residues on peptides, folding of proteins by various chaperones, the formation of disulfide bonds, the integration of proteins into membranes, the packaging of proteins into vesicles, and the turnover of misfolded proteins.

STAINING THE GOLGI APPARATUS

THE GOLGI APPARATUS

The Golgi apparatus is a modification and packaging organelle. The morphology of the Golgi apparatus is determined by its proximity to the endoplasmic reticulum. The cis face of the Golgi apparatus is closest to the endoplasmic reticulum, the medial face is in the middle, and the trans face is the furthest away. The Golgi apparatus is formed of cisternae similar to the rough endoplasmic reticulum; however, large membranous sacs are found in the cis-face of the Golgi apparatus where tubules are found in the trans-face of the Golgi apparatus. These tubules are often referred to as the trans-Golgi network.

From transcription to final processing, the path is as follows: transcription and processing of a messenger RNA in the nucleus, transport of the message to a ribosome on the rough endoplasmic reticulum or a free ribosome in the cytosol, modification and packaging within the rough endoplasmic reticulum and packaging into a transport vesicle, transport through the cis-, medial-, and trans-faces of the Golgi apparatus, final packaging into transport vesicles at the trans-Golgi network.

More complex glycosylation or addition of carbohydrates can be performed within the Golgi apparatus; it is also the site of more complex polysaccharide synthesis. There are two models for transport through the Golgi apparatus: the cisternal maturation model suggests that the Golgi apparatus matures from one face to the other starting at the cis-face and slowly maturing to become the trans-Golgi network. However recent experimentation leads us to believe that the vesicular transport model is more likely. Peptides are delivered to the cis-face of the Golgi apparatus and are transported through the cisternae via transport vesicles until they reach the trans-Golgi network where they are ultimately packaged and move on to their final destination somewhere else in the cell.

Today you will be observing the morphology of the endoplasmic reticulum and Golgi apparatus, using histological procedures specific for labeling of each organelle.

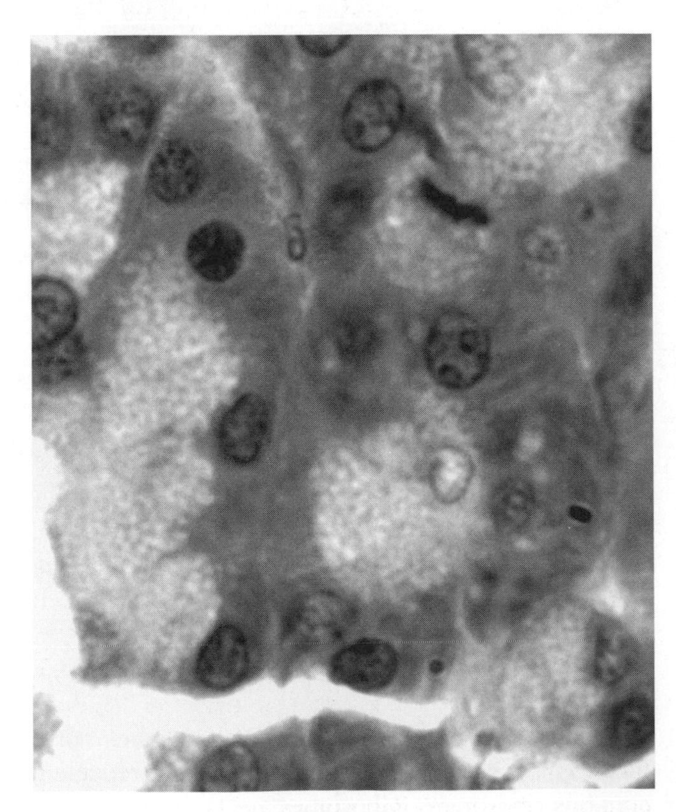

PROCEDURES

Staining of Intracellular Membranes and the Endoplasmic Reticulum

1. You will receive a 6-well plate with 3 wells filled with a mammalian cell culture. Label the first well "unstained," the second well "membrane," and the third well "ER." Move your plate to the cell culture hood.
2. Remove the medium from well #3 with your pipette and add 3 mL of ER-Tracker stock solution. Wrap the plate in aluminum foil, and incubate at 37°C for 20 minutes.
3. Pull your plate from the incubator, and take it to the culture hood. Remove the medium from well #2 and well #3 with your pipette and add 3 mL of GFP Counterstain (BODIPY® TR methyl ester) stock solution. Wrap the plate in aluminum foil, and incubate at 37°C for 10 minutes.
4. Pull your plate from the incubator, and take it to the culture hood. Remove the staining solution from well #2 with your pipette, and add 50 μL of Hanks' balanced salt solution containing 10 mM HEPES, pH 7.4.
5. Using a PAP pen, create three wells on a lysine-treated Probe-on-Plus slide. Label the wells 1, 2, and 3.
6. Transfer 50–100 μL of culture from the matching well on your culture plate to the appropriate well on the slide. Be careful to avoid exposing your sample to direct light.
7. Observe the cells in each well, using light and UV-Fluorescence microscopy. Answer questions 1 and 2 on the worksheet.

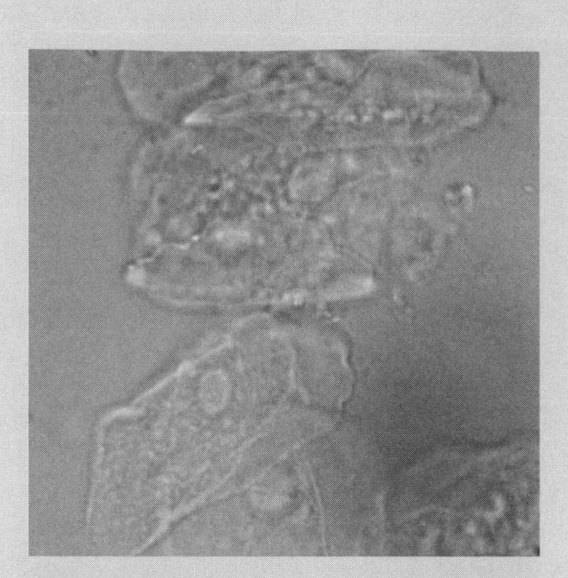

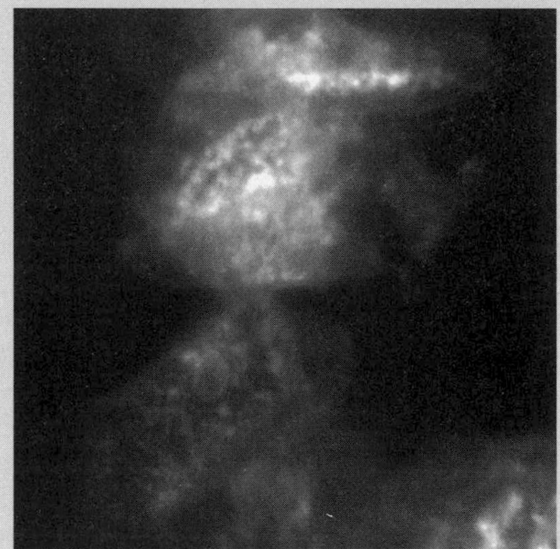

Fluorescent ER-Labeling

Observing Silver-Stained Pancreas Preparations

1. While your other samples are staining, observe one of the prepared slides that display a special silver-stained section of mammalian or amphibian pancreas. This preparation should highlight the Golgi apparatus. Silver-staining should appear as a dense, dark stain on the slide.
2. Use your observations to answer question 3 on the worksheet.

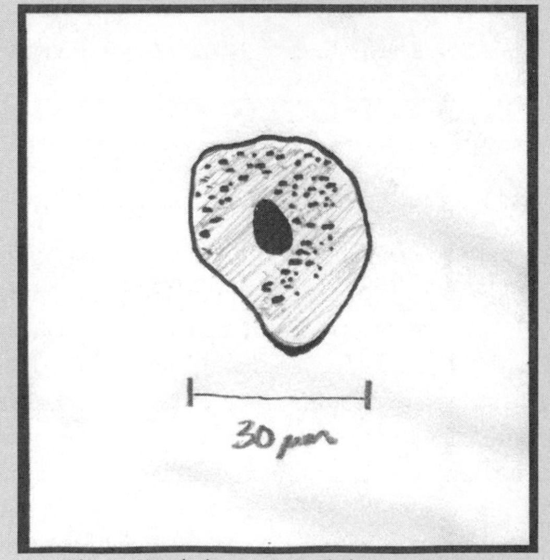

Silver-Staining of Golgi Apparatus

WORKSHEET

Observations and Assessments

1. What do the unstained cells look like? Describe any visible structures, and draw an example sketch in the space provided below. Feel free to use color in your sketch, if you prefer. What do the membrane-stained cells look like? Can you see the endoplasmic reticulum now? Can you see the Golgi apparatus? Can you see any transport or secretory vesicles? Describe any visible structures, and draw an example sketch in the space provided below. Remember to label all microscope images with the following: organism, cell type, magnification, and stain; add a scale bar at the bottom right of each sketch.

2. What do the ER-Tracker-stained cells look like? Can you distinguish between the endoplasmic reticulum and the other membranes now? Can you see differences in the transport vesicles? Describe any visible structures, and draw two example sketches comparing the membrane-staining and ER-staining in one cell the space provided on the following page.

3. What do the silver-stained pancreas cells look like? Can you see the Golgi apparatus? What is the silver staining, and how does this help you to visualize the Golgi apparatus? Describe any visible structures, and draw an example sketch in the space provided below. Draw an image of the Golgi apparatus observed under the high power objective, and label any structures that you recognize. Explain what is being stained by the silver in this histological preparation.

____

____

____

____

____

____

4. Using the Internet, look up the information for the ER-Tracker and GFP Counterstain (BODIPY® TR methyl ester) stains. Describe what each stain is labeling. Explain why they had to be used together, in combination, for this experiment.

5. Today we used light and fluorescence microscopy to identify two large organelles. Which staining method looked better? Which was more specific? What are the benefits of fluorescence microscopy? What are the shortcomings? Discuss when you should use light microscopy versus when you should use fluorescence microscopy.

DISCUSSION

Discussion Questions

1. While the membranes of the endoplasmic reticulum can be identified using fluorescent labeling, other aspects of the ultrastructure of this organelle required Transmission Electron Microscopy to visualize. Discuss the differences between UV-Fluorescence Microscopy and Transmission Electron Microscopy. From what you know about the structure of the ER, discuss what features would be visible using UV-Fluorescence Microscopy; then discuss what features would not be visible using UV-Fluorescence Microscopy.
2. Silver is used to stain the Golgi apparatus in the prepared slides. Discuss what is being stained by the silver, and why this is an excellent stain for the Golgi apparatus. Also, use the Internet to look into the discoveries of Camillo Golgi, and discuss why this organelle bears his name.

3. Chaperones are large protein complexes that assist in the proper folding of proteins in the cytoplasmic cytoplasm and ER. Chaperones may be involved in regular protein folding, or they may be expressed to assist with protein folding during stress conditions. Typical stresses that induce chaperone production include heat, cold, pH, oxidative stress, osmotic stress, and salt stress. Design an experiment that would induce the possible production of ER-chaperones, and explain how you might detect these chaperones in the ER.

4. Tunicamycin is an antibiotic that induces the unfolded protein response, leading to an accumulation of N-linked glycoproteins in the lumen of the ER. This leads to immediate ER-stress, and an eventual arrest of the cell cycle at G1-phase. Continuous ER-stress will typically lead to the induction of apoptosis and the death of the cell. Devise an experiment that might lead to a reduction of the stress caused by this antibiotic.

The Endocytic Pathway (FM4-64 Staining)

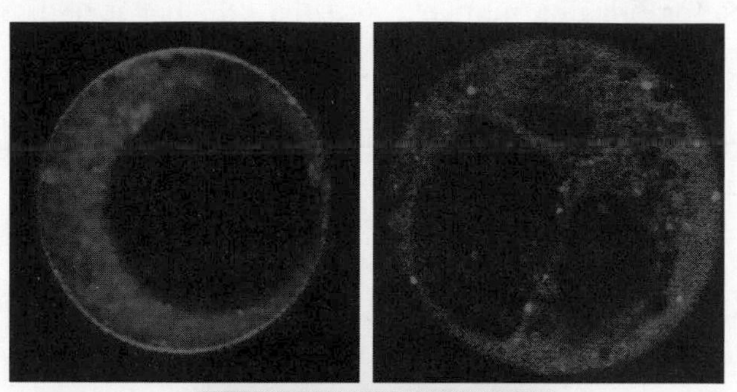

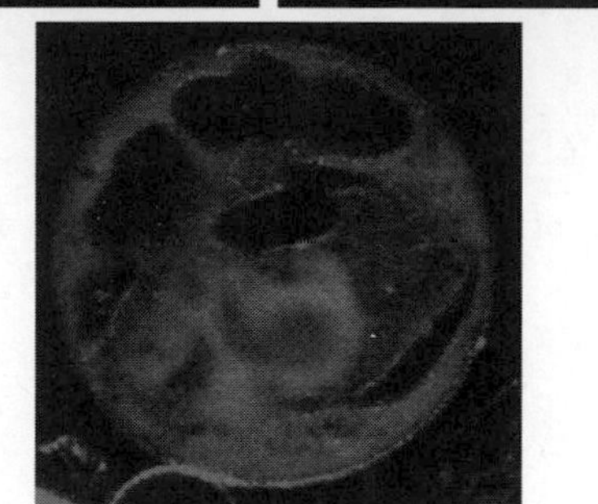

Objectives

- Prepare and observe plant protoplasts labeled with FM4-64 staining.
- Track the movement of the fluorescent FM4-64 stain as it is internalized.

24

OVERVIEW

DESCRIPTION

In order to survive, the cell must be able to absorb materials from the extracellular environment. Endocytosis is the absorption of extracellular compounds, either in bulk or bound to receptors on the exterior of the plasma membrane. There are two kinds of endocytosis: bulk phase endocytosis, which is also called pinocytosis, and receptor-mediated endocytosis. In receptor-mediated endocytosis, the receptors allow for specificity specific ligands or molecules of interest to bind to receptors on the plasma membrane and become internalized as vesicles that will mature as and form endosomes. There are two types of endosomes: early endosomes and late endosomes. In mammalian cells, late endosomes mature into lysosomes. In plant cells, endosomes are transported to the central vacuole. The material in the endosome is digested and released into the cytoplasm. This unwanted material is then released as the residual body fuses with the interior of the plasma membrane and releases its contents through exocytosis. The staining compound FM4-64 will bind to the plasma membrane and be internalized as it enters the cell. During this process, you will be able to see the movement of the stain from the exterior of the cell, on the plasma membrane, into both early and late endosomes and finally to its final resting place in the vacuole membrane. Thus you will be able to track the endocytic pathway in a plant cell for your analysis.

CONCEPTS & VOCABULARY

- Endocytosis
- Endosome
- Exocytosis
- Lysosome
- Phagocytosis
- Pinocytosis
- Tonoplast
- Vacuole

CURRENT APPLICATIONS

- FM4-64 is an extremely versatile stain for plant cells. It can be used to identify the plasma membrane, the endosomes, and the vacuolar membrane (tonoplast). When used in conjunction with other fluorescent stains, FM4-64 can be used to highlight specific membranes at different incubation times.
- Endocytosis may play an important role in the prevention of cancer. Not only is endocytosis essential for bringing nutrients into the cell, but it is also important for the recycling of receptors bound to the plasma membrane. If this recycling process is altered, it may lead to a proliferation of receptors and altered signaling/control of the cell cycle and growth.

REFERENCES

Karp, G. (2010). The endocytic pathway. *Cell and Molecular Biology: Concepts and Experiments*. (311–324). Hoboken, NJ: John Wiley & Sons, Inc.

Polo, S, Pece, S, Di Fiore, PP. (2004). Endocytosis and cancer. *Curr. Opin. Cell. Biol.* 16(2):156–61.

BACKGROUND

EXTRACELLULAR ABSORPTION

Just as a cell is able to secrete proteins and compounds from the interior to the exterior of the cell, in order to survive, the cell must also be able to absorb materials from the extracellular environment. Endocytosis is the absorption of extracellular compounds, either in bulk or bound to receptors on the exterior of the plasma membrane. There are two kinds of endocytosis: bulk phase endocytosis, which is also called pinocytosis and receptor-mediated endocytosis. Bulk phase endocytosis or pinocytosis is non-specific. Small portions of the extracellular fluid fill tiny vesicles on the exterior of the plasma membrane and enter the cytoplasm. This cell-drinking is used to absorb tiny portions of the extracellular fluids in order to gather bulk nutrients.

Receptor-mediated endocytosis is specific. Using specific receptors that bind to ligands of interest, compounds are accumulated and brought into the cell. Another form of receptor-mediated endocytosis is phagocytosis. Phagocytosis is used by the cells of the immune system and cells capable of amoeboid movement to engulf large amounts of particulate matter in order to digest nutrients, or in the case of immune cells foreign proteins or organisms. In this way endocytosis is serving a monitoring function for the immune system rather than the gathering of extracellular components.

ROLE OF RECEPTORS

In receptor-mediated endocytosis, the receptors allow for specificity; specific ligands or molecules of interest bind to receptors on the plasma membrane. These receptors are often found in pits that are coated by a compound called clathrin. These clathrin-coated pits are covered with receptors and allow for an increased surface area as they pucker inward toward the interior of the cell. Once the clathrin-coated pit has been filled, with all receptors bound to ligands, it will move inward and form a vesicle in the cytoplasm. These vesicles will mature as they lose their clathrin coating and form endosomes.

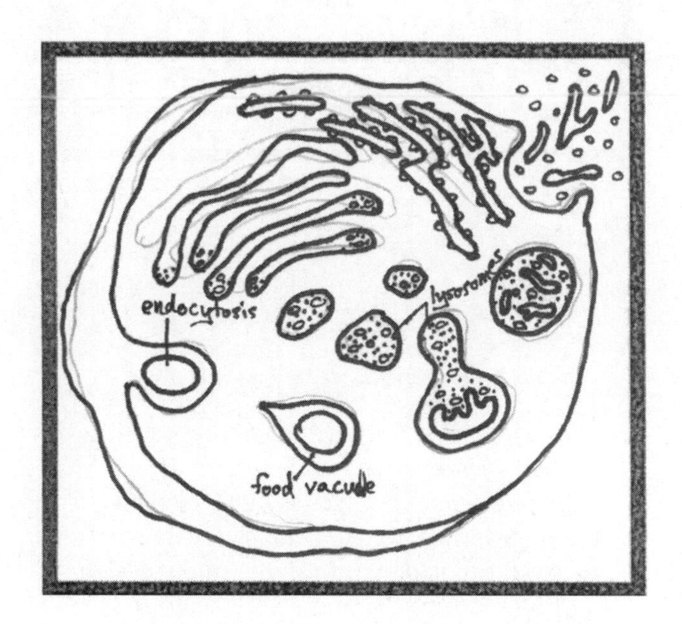

There are two types of endosomes, defined by their location in the cell. Early endosomes are typically found in the periphery, near the plasma membrane. Late endosomes are delivery vesicles and are found closer toward the interior of the cell, usually near the nucleus. Early endosomes have a higher pH than late endosomes, which are quite acidic. In mammalian cells, late endosomes mature into lysosomes as digestive enzymes are delivered to the endosome, converting it into a functional lysosome when the enzymes enter the acidic lumen of the late endosome. Other receptors control the endocytic pathway and are very important for the recycling of membranes, clathrin, and the external receptors that bind the ligands on the outside of the plasma membranes. This process would be less efficient if the components could not be recycled and used multiple times.

However, in plant cells, endosomes are transported to the central vacuole, since the central vacuole contains the low-pH environment and digestive enzymes required for breakdown of extracellular components. The late endosome membrane will fuse with the vacuole membrane or tonoplast, and the contents of the late endosome will be pushed into the interior of the vacuole. These contents will then be degraded by the enzymes found therein.

During phagocytosis, portions of extracellular material are engulfed and enter the cell to form a phagosome. This phagosome will bind to a lysosome to form a phagolysosome. The digestive enzymes within the lysosome will break down the contents of the phagosome, and permeases will release those digestive components to the cytoplasm. Anything that is not released to the cell by permeases is kept as the phagosome becomes a membrane-bound structure called a residual body. This unwanted material is then released as the residual body fuses with the interior of the plasma membrane and releases its contents through exocytosis.

In immune cells, bacteria are often killed by hydrogen peroxide or other reactive oxygen species found within the phagosome. This prevents the bacteria from using the phagosome as an entry point into the cell. However, this is not the case for all bacteria. Some bacteria are able to deactivate this killing mechanism and gain access to infect the cell through phagocytosis.

LYSOSOMES AND VACUOLES

In mammalian cells, late endosomes will merge with vesicles from the trans-Golgi network containing lysosomal enzymes. Lysosomes are essentially multiple tiny stomachs found in the cell used for the degradation of materials. This degradation can be used to gather nutrients for intracellular processes or can be used to recycle used, damaged, or unneeded components made by the cell.

In plant cells, however, the late endosome is delivered to the central vacuole, and the contents are digested. These two processes while ending in a similar result use different systems, both of delivery and of maturation of the endosome. This will be important today since we are observing plant cells and their endocytic pathway rather than mammalian cells.

The staining compound FM4-64 will bind to the plasma membrane and be internalized, and as it enters the cell, it will piggy-back on the membranes of the endosomes. The early endosomes typically start out small, and then they get slightly larger as they become late endosomes. The late endosomes will move to the vacuole membrane or tonoplast, and the FM4-64 finally will label the tonoplast membrane itself. During this process, you will be able to see the movement of the stain from the exterior of the cell, on the plasma membrane, into both early and late endosomes, and finally to its final resting place in the vacuole membrane, using the florescence of the compound itself and UV florescence microscopy. Thus, you will be able to track the endocytic pathway in a plant cell for your analysis.

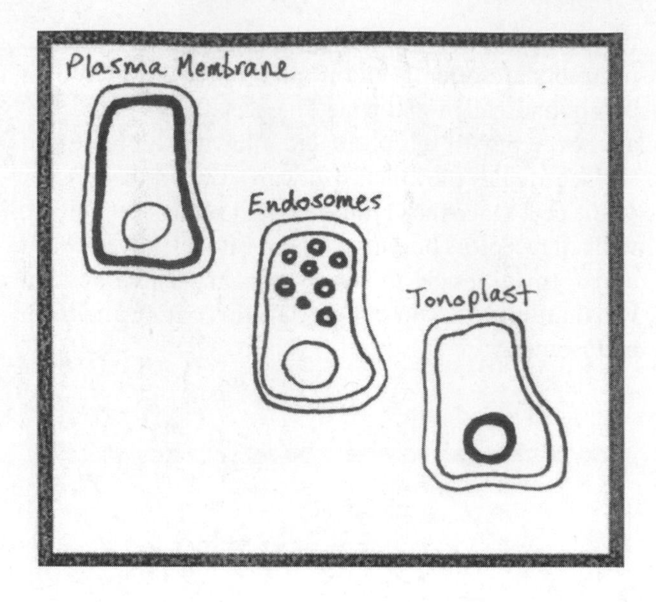

PROCEDURES

FM4-64-Staining of protoplasts

1. Keep cells in medium on ice before you begin the experiment.
2. Add 6.0 μL of FM4-64 stock to your sample. Wrap your culture plates/tubes in aluminum foil to prevent light exposure.
3. Incubate for 10–20 min on ice.
4. Spin cells down and wash in medium.
5. Repeat step 4.
6. Using a PAP pen, create one well on a lysine-treated Probe-on-Plus slide.
7. Transfer 50–100 μL of culture from the matching well on your culture plate to the appropriate well on the slide. Be careful to avoid exposing your sample to direct light.
8. Observe your cells immediately using the UV-Fluorescence Microscope. Answer question 1 on your worksheet.
9. Wait 1 hour for FM4-64 to fully enter the endosomes. Using a PAP pen, create one well on a lysine-treated Probe-on-Plus slide.
10. Transfer 50–100 μL of culture from the matching well on your culture plate to the appropriate well on the slide. Be careful to avoid exposing your sample to direct light.
11. Observe your cells, using the UV-Fluorescence Microscope. Answer question 2 on your worksheet.
12. Get a stained sample that has been incubating for 4 hours from your instructor.
13. Using a PAP pen, create one well on a lysine-treated Probe-on-Plus slide.
14. Transfer 50–100 μL of culture from the matching well on your culture plate to the appropriate well on the slide. Be careful to avoid exposing your sample to direct light.
15. Observe your cells using the UV-Fluorescence Microscope. Answer question 3 on your worksheet.

WORKSHEET

Observations and Assessments

1. What do the unstained protoplasts look like? Describe any visible structures, and draw an example sketch in the space provided below. Feel free to use color in your sketch, if you prefer. What do the FM4-64 stained protoplasts look like immediately after staining? Can you see the plasma membrane fluorescing now? Can you see any other structures? Describe any visible structures, and draw an example sketch in the space provided below. Remember to label all microscope images with the following: organism, cell type, magnification, and stain; add a scale bar at the bottom right of each sketch.

____

____

____

____

____

____

2. What do the FM4-64 stained protoplasts look like 1 hour after staining? Can you see the plasma membrane fluorescing now? Can you see endosomes? Count how many endosomes are close to the plasma membrane (early endosomes) and how many are distant from the plasma membrane (late endosomes). Record your data for 5 cells. Can you see any other structures? Describe any visible structures and draw two example sketches in the space provided below. Remember to label all microscope images with the following: organism, cell type, magnification, and stain; add a scale bar at the bottom right of each sketch.

____

____

____

____

____

____

24

OVERVIEW

3. What do the FM4-64 stained protoplasts look like 4 hours after staining? Can you see the plasma membrane fluorescing now? Can you see endosomes? Count how many endosomes are close to the plasma membrane (early endosomes) and how many are distant from the plasma membrane (late endosomes). Record your data for 5 cells. Can you see the vacuolar membrane (tonoplast)? Can you see any other structures? Describe any visible structures, and draw two example sketches in the space provided below. Remember to label all microscope images with the following: organism, cell type, magnification, and stain; add a scale bar at the bottom right of each sketch.

4. Describe the difference in endosome labeling between each of your 3 time points. When did you see the most endosomes labeled? Was there a difference between early- and late-endosome labeling in any of the samples? If you were going to design an experiment that required labeling of late endosomes, how could FM4-64 be used? Give a brief description of the staining protocol you would use to label late endosomes.

____

____

____

____

____

____

5. Using the space below, draw a diagram of the internalization and passage of the FM4-64 dye through the protoplast. Use arrows to indicate direction of flow, and label each labeled membrane with a time point when they should be best labeled.

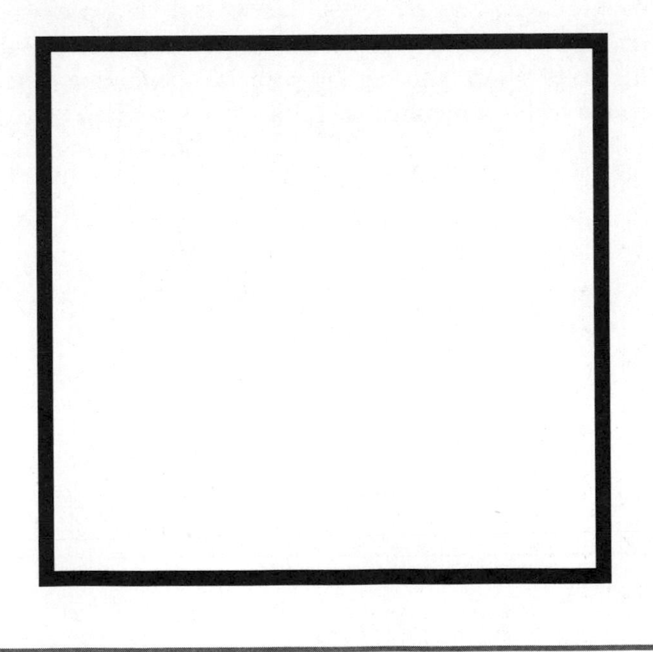

DISCUSSION

Discussion Questions

1. Discuss the action of FM4-64. Use the Internet to determine how it labels membranes and how this action can be used to label the endocytic pathway.
2. Discuss the properties of early endosomes. Discuss the properties of late endosomes. What is the source of the early endosome membrane? What is the source of the late endosome membrane?
3. What role does the lysosome play in the endocytic cycle of mammalian cells? What role does the vacuole play in the endocytic cycle of plant cells? Describe the differences between these two organelles.
4. Define pinocytosis and phagocytosis in terms of their similarities and differences. How does phagocytosis differ from macroautophagy? How are the two processes similar?
5. Discuss your expected results if you stain mammalian cells with FM4-64. How would they differ from what you saw today? Why were protoplasts chosen for this experiment?

Detection of MAP Kinase Signaling

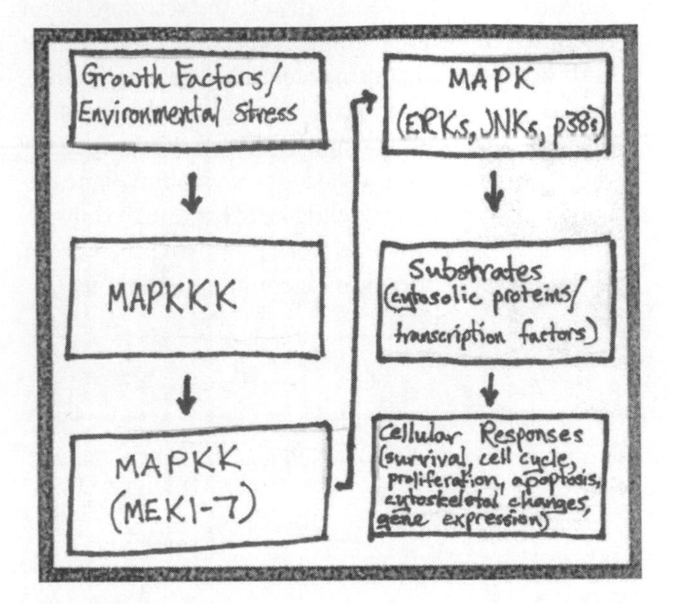

Objectives

- Observe the activity of a purified MAP Kinase.
- Prepare a protein extract and observe MAP Kinase activity.

DESCRIPTION

Cell signaling is a phenomenon in which information can be relayed across the plasma membrane into the interior of the cell and eventually to the cell nucleus. This process can also be referred to as signal transduction. The final target of the majority of signals is either the activation of a protein, typically through phosphorylation, or the activation of a transcription factor, again through phosphorylation, which leads to a change in gene expression. There are many signals in which the first stimuli to initiate the cascade binds to a G protein-coupled receptor (GPCR), which uses the energy from GTP hydrolysis for activation of the next receptor found within the interior side of the cell membrane. This event is usually a phosphorylation event. Other signals may bind to the extracellular domain of a receptor protein tyrosine kinase (RTK). This activation is similar to the activation of G proteins, but instead a tyrosine kinase domain is activated on the inner surface of the plasma membrane. Kinases are proteins involved in the phosphorylation of other proteins. A good example of one of these signaling pathways is the MAP Kinase Cascade. The MAP Kinase Cascade involves a GTP binding protein called *Ras*. When activated by an RTK, *Ras* will bind to another protein called *Raf*, which will be recruited to the plasma membrane and form an active protein kinase. This protein kinase will activate the first MAP kinase in a cascade. A MAP kinase kinase kinase (MAPKKK) is activated through phosphorylation, which will then phosphorylate a MAP kinase kinase (MAPKK), which will then phosphorylate a MAP kinase (MAPK), which will either lead to a change in gene expression. In the exercise you are performing today, you will be measuring the activity of various MAP kinases.

CONCEPTS & VOCABULARY

- Cell signaling
- G-Protein
- G-Protein-coupled receptor
- MAP Kinase (MAPK)
- Phosphorylation, *Ras/Raf*
- Receptor Protein Tyrosine Kinase (RTK)
- Signaling cascade

CURRENT APPLICATIONS

- One MAPKKK can activate one or more MAPKK enzymes. Each of those MAPKKs may activate one or more MAPKs, each leading to a specific change in gene expression and response. As you might imagine, the pathways involving MAPKs become quite complex, as there are hundreds of identified MAP kinases in the human genome. This topic is a rich source of current molecular research.
- Kinase assays are used to dissect the response of an organism to various stimuli. The sum of all kinase pathways in an organism is referred to as the *kinome*. Just as there are large-scale projects for the genomes of various organisms, kinome projects are the focus of very active research, especially in mammalian and human tissues. If you would like to explore the human kinome (or order a free poster showing all known pathways) visit http://kinase.com/human/kinome/.

REFERENCE

Karp, G. (2010). Protein-tyrosine phosphorylation as a mechanism for signal transduction. *Cell and Molecular Biology: Concepts and Experiments.* (634–641). Hoboken, NJ: John Wiley & Sons, Inc.

BACKGROUND

CELL SIGNALING

Cell signaling is a phenomenon in which information can be relayed across the plasma membrane into the interior of the cell and eventually to the cell nucleus. Typically, cell signaling involves the recognition of some signaling molecule at the plasma membrane on the outer surface. This signal causes a conformational change in a bound receptor protein, and the conformational change allows for the transduction of the signal across the plasma membrane. This process can be referred to also as signal transduction. Once the signal has been transduced through the membrane, other proteins associated with the receptor will be modified and thus transfer the signal to the interior of the cell.

Several different components of the signaling cascade may be involved until the final target of the signal is reached. The final target of the majority of signals is either the activation of a protein, typically through phosphorylation, or the activation of a transcription factor, again through phosphorylation, which leads to a change in gene expression.

G PROTEINS

There are many signals in which the first stimuli to initiate the cascade binds to a G protein-coupled receptor (GPCR). G proteins can be found on the outer surface of the cell membrane and are transduced to release a second messenger inside the cell. These proteins are referred to as G proteins because they specifically bind to guanine nucleotides, either GDP or GTP. The G protein can exist in two possible states, either an active state, when it is bound to GTP, or an inactive state, when it is bound to GDP. The G protein–coupled receptors are a large family with each protein responding to a specific stimuli. They all act through a similar mechanism. The signal molecule will bind to a receptor and cause a change in its conformation that allows for the binding of a G protein. The G protein will bind and release its GDP, and bind an available GTP, which will shift the G protein into an active state.

After activation, the GDP will be hydrolyzed and converted into GTP, which will deactivate the protein, but allow for the transduction of the signal to the next receptor in the cascade. The energy from GTP hydrolysis is used for activation of the next receptor typically found either on the interior side of the cell membrane or within the cytoplasm itself. This event is usually a phosphorylation event.

RECEPTOR PROTEIN TYROSINE KINASE AND THE MAP KINASE

Other extra cellular signals may bind to the extra cellular domain of a receptor protein tyrosine kinase (RTK). This activation is similar to the activation of G proteins, but instead a tyrosine kinase domain is activated on the inner surface of the plasma membrane. Kinases are proteins involved in the phosphorylation of other proteins. Phosphorylation of a protein will cause a conformational change; this may activate or deactivate the protein. In the case of cellular signaling, an activation will typically involve another kinase being activated and thus activating an entire signaling pathway. A good example of one of these signaling pathways is the MAP Kinase Cascade.

The MAP Kinase Cascade involves a GTP binding protein called Ras. Like other G proteins, Ras cycles between the inactive GDP form and the active GTP form. When Ras is activated, it will stimulate the activation of other effector molecules downstream in the signaling pathway. When a ligand binds to the RTK, the RTK is autophosphorylated, and the cytoplasmic domain of the receptor will recruit a protein called SOS, which is the activator of Ras, found on the inner surface of the cell membrane. Ras will catalyze the exchange of GDP for GTP and activate the Ras protein. The activated Ras will bind to another protein called Raf, which will be recruited to the plasma membrane and form an active protein kinase. This protein kinase will activate the first MAP kinase in the cascade.

A MAP kinase kinase kinase is activated through phosphorylation, which will then phosphorylate a MAP kinase kinase, the MAP kinase kinase will phosphorylate a MAP kinase, which will either lead to a change in gene expression or activate a final transcription factor. That transcription factor will bind to the promoter region of the gene and either activate or deactivate its expression.

In the exercise you are performing today, you will be measuring the activity of various MAP kinases in a MAP kinase cascade. The assay will measure the phosphorylation of myelin basic protein (MBP) by the kinase(s) in a protein sample and then detect the phosphorylated-MBP by Western Immunoblot detection. You will be using kinases extracted from human cells, using a purified human MAP Kinase 2/Erk2 as a positive control.

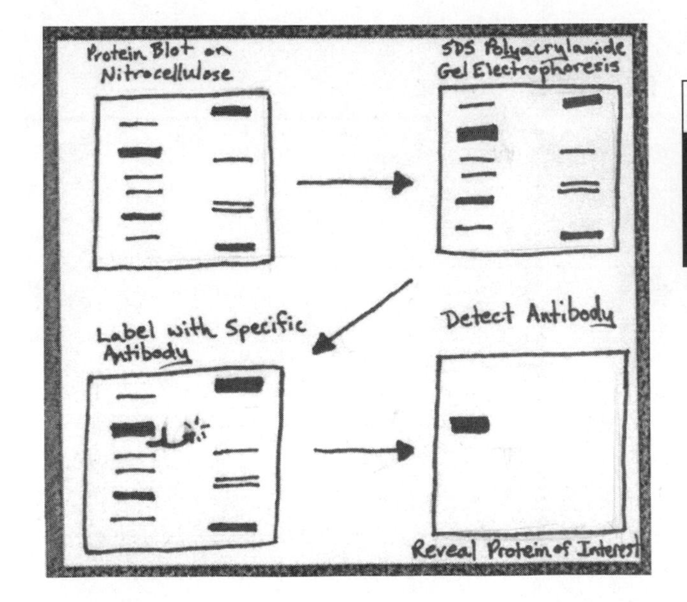

25 OVERVIEW

PROCEDURES

Preparation of the Total Protein Extract

1. Obtain a mammalian suspension cell culture from your instructor and take it to the culture hood.
2. Transfer 1.0 mL of culture to a 1.5 mL microcentrifuge tube.
3. Spin the cells down at 2500×g for 10 minutes. Discard the supernatant.
4. Wash the cell with 1.0 mL of 1X PBS, and spin the cells down at 2500×g for 10 minutes. Discard the supernatant.
5. Add 300 μL M-PER reagent to the cell pellet. Pipette the mixture up and down to resuspend pellet.
6. Shake mixture gently for 10 minutes.
7. Remove cell debris by centrifugation at 14,000×g for 15 minutes.
8. Transfer the supernatant to a new tube for analysis.

Assay of Protein Kinase Activity

1. Keep all solutions and samples on ice until otherwise instructed.
2. Mix 10 μL of substrate cocktail, 10 μL of inhibitor cocktail, and 10 μL of protein extract preparation into the bottom of a microcentrifuge tube. This is your experimental sample.
3. Mix 10 μL of substrate cocktail, 10 μL of inhibitor cocktail, and 10 μL of MAP kinase 2/Erk2 control sample into the bottom of a microcentrifuge tube. This is your positive control.
4. Start the reaction by adding 10 μL of the Mg2+/ATP cocktail, vortex gently, and incubate the microcentrifuge tube shaking at 30°C for 30 minutes.
5. These samples can now be stored or used immediately for Western Immunoblot Analysis, using anti-Phospho-MBP antibodies as a primary antibody.
6. If using immediately, transfer 2.5 ml of the reaction mixture into another microcentrifuge tube. Add 7.5 ml of 1× TBS and 10 ml of 2× Laemmli sample buffer. This sample is now ready to be loaded onto an SDS-PAGE gel.
7. Use the protocol in the SDS-PAGE and Western blotting chapters to continue.

WORKSHEET

Observations and Assessments

1. What cell/tissue did you use for your experiment today? Do you expect strong or weak kinase activity in this sample? What might be done to increase the kinase activity? Explain your answers.

2. What positive control did you use today? What negative control was used? Describe the level of kinase activity that you expect in these samples, compared to your protein extract sample.

3. Paste a photo or draw your Western Immunoblot results in the space below. Describe the kinase activity seen in your protein extract samples. Compare that activity with your positive and negative controls. Do your results meet your expectations from questions 1 and 2?

4. What is being phosphorylated by the kinases in your sample and positive control? What are you measuring with your Western Immunoblot Assay? Is this experiment detecting specific kinases or total kinase activity? What might be done to increase the specificity of this assay?

5. This experiment did not use radioactive phosphorous as a marker for kinase activity. How is radioactive phosphorous used to detect kinase targets? Discuss the pros and cons of using radiolabeling in a kinase assay.

DISCUSSION

Discussion Questions

1. Discuss how the phosphorylation or dephosphorylation of a protein can cause a change in its activity. How might this be used in a kinase cascade?
2. What is a G-protein–coupled receptor? Where are these receptors located? How do they translate extracellular signals into intracellular responses?
3. Discuss how receptor tyrosine protein kinases (RTK) and MAP kinases work together to produce a change in gene expression.
4. You have discovered that a specific kinase increases in expression in skin cells during exposure to sunlight. You have also discovered that a mutation in this kinase may lead to skin cancer. Do some research, and design an experiment that would reveal the targets of this kinase.

Cancer

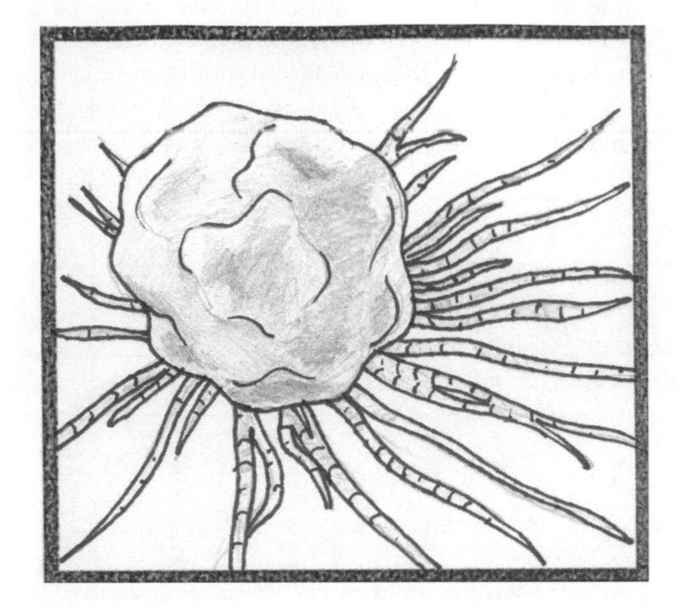

Objectives

- Read and understand the contents of Chapter 16 in Karp.
- Discuss topics involved in cancer research.
- Learn more about specific causes of cancer cell formation.

26 OVERVIEW

DESCRIPTION

Cancer is a condition in a cell when uncontrolled growth and cell division, de-differentiation, and immunity from cell death are the norm. Cancer has been linked to mutations in a number of genes, but no one mutation can cause cancer. Mutation in at least two genes is required to form a cancer cell: an oncogene, which regulates cell proliferation; and a tumor-suppression gene, which regulates the cell cycle, monitors DNA damage in cells, and regulates apoptosis. These mutations can be hereditary, but the majority of cancers have sporadic origins. Anything that increases the rate of mutation, however, can lead to an increased cancer risk. Cancer cells have a number of properties that make them unique to other types of aberrant cells. These properties include immunity to growth signals, unchecked growth, loss of apoptosis and senescence, the ability to induce angiogenesis, and the ability to invade neighboring tissues. Most cancer treatments attack one of these properties to deactivate the cancer and kill the malignant cells.

CONCEPTS & VOCABULARY

- Aneuploidy
- Apoptosis and programmed cell death
- De-differentiation
- DNA repair gene
- Metastasis
- Oncogene
- p53
- Senescence
- Tumor-suppressor gene

CURRENT APPLICATIONS

- Since most chemotherapy involves inducing cell death or preventing DNA synthesis and/or cell division, they can have detrimental effects on all cells in the body. This is why much research is focused on cancer therapies that are localized or targeted act on tumor cells.
- Research at the University of Alberta in the laboratory of Dr. Evangelos Michalakis has found that the drug dichloroacetate was effective at reducing tumor size in rats by restoring mitochondrial function. While these early results are promising, DCA is still only in Phase II Clinical Trials for human treatments.

REFERENCES

Futreal, PA, et al. (2004). A census of human cancer genes. *Nat Rev Cancer*. 4:177–83.

Gibbs, WW. (2003). Untangling the roots of cancer. *Scientific American*. 289(11), 48–57.

Karp, G. (2010). Cancer. *Cell and Molecular Biology: Concepts and Experiments.* (650–681). Hoboken, NJ: John Wiley & Sons, Inc.

PROCEDURES

The Cancer Board Game

The purpose of this game is to illustrate the factors involved in the transformation of one cell into a cancerous cell. Most cancers are the end result of a series of mutations of very specific genes in a single cell. Alfred Knudson proposed the two-hit hypothesis of tumor formation in 1971 as he studied retinoblastoma. Knudson suggested that in order for a tumor to form in his patients, a mutation in at least two genes was required: the RB tumor suppression gene and a separate oncogene. A mutation in only one of these genes would not transform a healthy cell into a cancerous cell. However, two mutations (or "two hits") were sufficient for retinoblastoma formation.

In this game, you will discover how events and lifestyle choices throughout a normal human life can lead to the accumulation of mutations in a single cell and potentially cancer. The game ends when all players have made one circuit around the game board. If a player has accumulated two harmful mutations by the end of the game, which we will assume that this leads to the formation of a cancer cell.

Gameplay (for 2 to 4 players)

The game begins with a die roll to determine the order of play. The person with the highest roll goes first. At the beginning of the game, each player will draw a birth card. Often the circumstances of our birth, along with our genetics, have a profound effect on our lifetime health. Each of the birth cards has specific rules that will remain in play for the entire game.

Players will roll a six-sided die to determine the number of spaces that they travel during each turn. Players will move their game pieces around the board, following the directions on each game space. The available game spaces include Educational events (purple), Occupational events (magenta), Stressful events (red), and Life events (yellow). Each game space will provide a draw or discard of a mutation card, a lifestyle card, or a money chip.

Mutation cards are kept by the player and laid out face-up in front of them for everyone to see. They may be neutral or harmful mutations. Lifestyle cards are read, and the instructions on the card should be followed. Lifestyle cards may be beneficial or harmful with 50–50 odds.

Money chips can be paid out to the bank (for taxes, hospital fees, or other costs). At the end of each turn, a money chip can also be used to buy off mutations (one chip per card) or to purchase a lifestyle card (one chip per card). If you wish to buy off a mutation card, you will need to advance/retreat to the hospital, spend one money chip, and forfeit your next turn. Only one mutation may be bought off during the game, and mutations may not be bought off if the player has already reached the end of the game board.

As you progress through the game, take the time to discuss when certain events or game cards can cause mutations. Also, discuss the differences between mutations and cancer causing mutations. Try to determine if there are any common factors in a person's life that lead to a predisposition to tumor cell formation.

DISCUSSION

Discussion Assignment

To prepare for this class, you must read Chapter 16 of Gerald Karp's *Cell and Molecular Biology* textbook before coming to class. There are a number of other helpful cancer review articles that can also be found in the reference section at the beginning of this chapter. Reading these articles is optional, but highly suggested. During class, you will play an educational board game in groups of four. Then, you will be required to answer one of the 14 discussion questions on the next page. Most answers to each of these questions can be found in the Karp text. However, even more detailed information can be found in the optional articles. Your instructor may also ask you additional questions based on the discussion. Students in the class may also ask you appropriate questions based on the discussion. Your response to your question will be worth 20 points, calculated using the rubric below.

26

OVERVIEW

Grading Rubric

Performance	Points
Completely answered all parts question	10
Preparation	5
Presentation	5
Added detail to responses	1–5 extra
Responses to additional questions from TA/classmates	1–5 extra
Added to the discussion/asked a good question	1–5 extra

Discussion Questions

1. Define cancer in terms of the cell cycle and growth. List the causes of cancer, along with the possible carcinogenic agents associated with each cause.
2. Define tumor-suppression gene and oncogene. Describe how they each lead to the formation of a cancer cell.
3. In the cancer game, what events or choices helped to remove mutations/prevent cancer? What events or choices led to the formation of new mutations/cancer? Discuss why these events and choices have this suggested effect on cancer and tumor formation in humans.
4. The most common cancers in the United States are lung, colon, breast, and prostate. What do the sites and cells of lung and colon cancers have in common? Why do you think that these cancers are so prevalent?
5. There are two main types of cancer: hereditary and sporadic. How frequently is cancer hereditary? Discuss in detail the main possible cause(s) of hereditary cancers. Is there any possible prevention for this form of the disease?
6. There are two main types of cancer: hereditary and sporadic. Discuss in detail the main possible cause(s) of sporadic cancers. Is there any possible prevention for this form of the disease?
7. What is a DNA-repair gene? How are mutations in these genes thought to lead to cancer cell formation? How do mutations in "master" genes that control cell division lead to cancer cell formation?
8. Define aneuploidy. Describe the all-aneuploidy theory of cancer formation.
9. Define apoptosis and angiogenesis. How are these two processes essential for tumor formation?
10. The p53 pathway is essential to our understanding of cancer. What does the p53 gene do, and why does a loss of p53 activity increase a patient's risk for cancer?
11. Genetic testing is available now for a number of cancer genes. Many people often view a positive test result as a virtual death sentence, but is this the case? Does the presence of a harmful mutation in an oncogene always lead to cancer formation? Does the presence of a harmful mutation in a tumor-suppressor gene always lead to cancer formation? Does the presence of a harmful mutation in the p53 gene always lead to cancer formation?
12. Cancer proteins normally regulate cell proliferation, cell differentiation, and cell death or mediate DNA-repair processes. Explain how each of these processes can lead to cancer cell formation.
13. Genes that control the cell cycle are important in preventing the formation of cancer cells. Describe what phases of the cell cycle might be changed in a cancer cell and explain why.
14. What are some current targets for cancer cell therapy? What do they target? How does this prevent the growth of tumors?
15. What side-effects are common to cancer treatments? If you were going to design a new cancer therapy, what cancer properties would you focus on and why?

Western Blot

Objective

- Perform hybridizations of a Western blot with primary and secondary antibodies.

DESCRIPTION

SDS-polyacrilamide gel electrophoresis (SDS-PAGE) is a technique that can be used to separate crude protein samples by mass. SDS is used to denature every protein in a sample, and the denatured sample is run on a polyacrilamide gel to separate each individual kind of protein. SDS-PAGE gels can also be used to create Western blots, which use antibodies to detect specific proteins based on structural elements that are unique to each protein, called *antigens*. In order to create a Western blot, the proteins separated in an SDS-PAGE gel must be transferred to a nitrocellulose membrane. This transfer can be accomplished by capillary action or electrophoretic mobilization. After the transfer, your protein bands will now be adhered to then membrane in the exact position where they were found on the gel. This is because the nitrocellulose paper is attractive to the negatively-charged, SDS-coated proteins. To prevent non-specific binding of antibodies, a blocking step is used. Blocking solution contains simple proteins that will not react with the antibodies. Most labs use powdered milk or bovine serum albumin as a common blocking agent. Then, the membrane is washed with solutions of primary and secondary antibodies. The primary antibody is specific to your protein-of-interest. The secondary antibodies are specific to the immunoglobin protein of the primary antibodies. Secondary antibodies are conjugated to some marker, which will later be used for detection. This is typically either an enzyme that creates light or a colored product that can be detected and recorded. Today, you will be hybridizing primary and secondary antibodies to a Western blot from your meat sample SDS-PAGE gel. You will then determine the presence or absence of soy proteins in your experimental meat samples and controls.

CONCEPTS & VOCABULARY

- Antigen
- Blocking agent
- ELISA
- Hybridization
- Nitrocellulose paper
- Ponceau solution
- Primary antibody
- Secondary antibody

CURRENT APPLICATIONS

- Western blotting is a powerful tool in the molecular laboratory. Where gene expression analysis can tell you if a gene is being activated, a Western blot can tell you how much protein is being made by the cell or tissue. Only the protein can effect the cell.
- While Western blotting is the preferred initial technique for research laboratory-based protein detection, the ELISA test is used more outside of the lab. Pregnancy and Rapid Strep Throat tests are two common examples. You can learn more about the ELISA test in the next online experiment.

REFERENCE

Karp, G. (2010). Isolation, purification and fractionation of proteins; the use of antibodies. *Cell and Molecular Biology: Concepts and Experiments.* (734–740; 763–765). Hoboken, NJ: John Wiley & Sons, Inc.

PROCEDURES

Blocking

- This step was done earlier today for 1 hour. This treatment will prevent antibody binding to non-specific protein binding sites on the membrane.

Primary Antibody

1. Pour off the wash solution, and add Primary Antibody solution.
2. Incubate the blot for 40 minutes at room temperature on an orbital shaker.
3. Pour the primary antibody solution back into the tube that it came in.
4. Then rinse in TBS-TWEEN for 3 minutes on the orbital shaker.
5. Repeat this rinse 2 more times. The wash solution can be discarded down the drain.

Secondary Antibody

1. Pour off the wash solution, and add Secondary Antibody solution.
2. Incubate the blot for 40 minutes at room temperature on an orbital shaker.
3. Pour the secondary antibody solution down the sink.
4. Then rinse in TBS-TWEEN for 3 minutes on the orbital shaker.
5. Repeat this rinse 2 more times. The wash solution can be discarded down the drain.

- Note: The following steps must be done quickly – so be prepared!

Detection

1. Remove the membrane from the wash solution, and dry for 1 minute.
2. Place the dry membrane into a clean plastic tray containing the detection solution. Your instructor will add detection solution BEFORE you add your membrane.
3. Place the tray on the shaker.
4. Watch the development of the bands (approximately 10 minutes or longer).
5. Pour off the detection solution, and rinse the blot with distilled water.
6. Photograph the developed blot.
7. Leave the blot to dry until next class meeting.

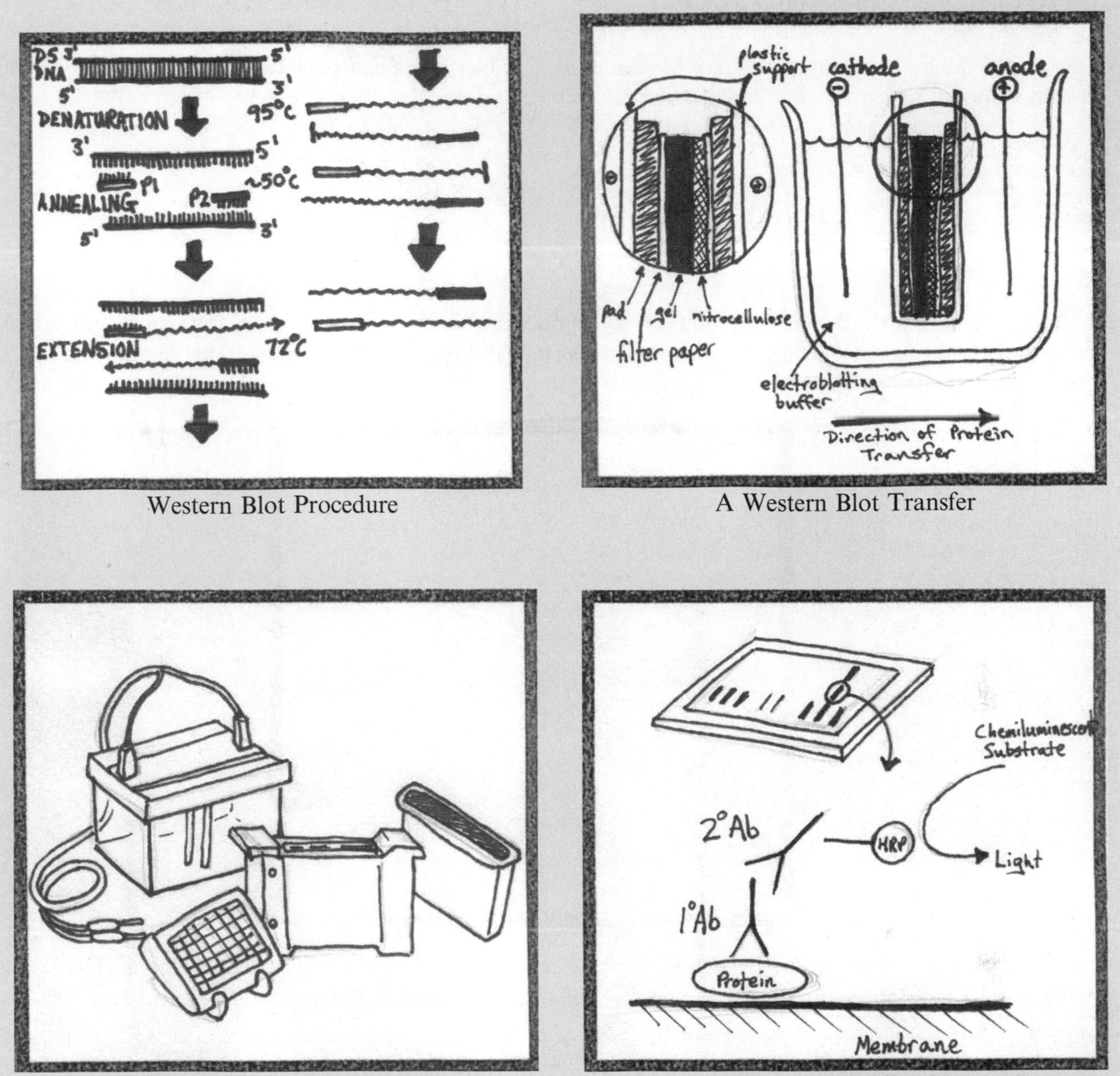

Western Blot Procedure

A Western Blot Transfer

A Western Blot Transfer Unit

Immunohybridization

WORKSHEET

1. What would your developed Western blot analysis look like if you had skipped the blocking step? Explain your response in terms of the importance of the blocking step.

2. Draw a picture (or tape a photograph) of your Ponceau-stained blot below. Label the lanes and samples for each. How does this compare with your stained SDS-PAGE gel? Be sure to label any important differences and similarities between the two.

3. Use the table below to compare the SDS-PAGE and Western blot results throughout the class. Detail MW-values for similar and dissimilar bands found in experimental versus control and at least one other set of protein samples.

Sample	Similar bands	Unique bands	Antibody hybridizations	Potential ingredients

4. You have developed your Western blot and found a faint band where the antibodies hybridized. What would you do to optimize your technique to increase the signal of the hybridized band? In another Western blot, you have a strong signal coming from one band, but weaker signal coming from several other bands that you think are nonspecific. How would you increase the specificity of your technique to determine if the weaker bands are specific or not?

5. Using the Western blot techniques that you have employed today, create a simple experimental design that describes how you would test the meat and fish products coming into a new Midwestern restaurant. Be sure to explain what you will use for samples, what you will use for controls, which methods you will use, and describe expected results for authentic versus imitation meat products. How does this design differ from your SDS-PAGE–based experimental design? Which is simpler, cheaper, and easier to understand? Which is more specific?

DISCUSSION

Discussion Questions

1. At a nearby imaginary school, a high school student nearly died after eating a hamburger. The hamburger was tested for bacterial contamination and was found to be negative for bacteria. No one else became ill after eating hamburgers from the same batch. The student has a severe allergy to soybean products. Soy products contain some of the proteins found in soy beans. Explain how a forensic scientist might analyze the hamburger meat to determine what caused the attack. Design an experiment similar to today's method that would test tofu and soy sauce for the presence of soy proteins. Decide what you would use for a control in each situation. Discuss any possible pitfalls with sample preparation and gel loading. What kinds of antibodies would you choose for the presence of soy?
2. Could protein isolation and an SDS-PAGE gel be used to confirm genetic identity in siblings, or paternity? How about Western blotting and immunodetection? Explain why, or why not.
3. You are working on a Western blot using an antibody that is specific for a human protein found in most cells. You wish to determine if there is an increase or decrease of this protein in blood samples in the presence of a certain drug. Can you identify any possible sources of contamination in your samples? Describe what steps you would have to take to avoid possible contamination.
4. Western blotting determines the presence of a protein. Is this proof of a protein's activity? Discuss why, or why not. Which is the more accurate test of identity: a Western blot/ELISA or an activity assay?
5. The simplicity of ELISA-based dipstick tests make them much more attractive to healthcare technicians and consumers at home. Can you think of a situation when a Western blot analysis would be preferable to an ELISA test? What factors would make you choose one technique over the other?

Enzyme-Linked Immunosorbent Assay

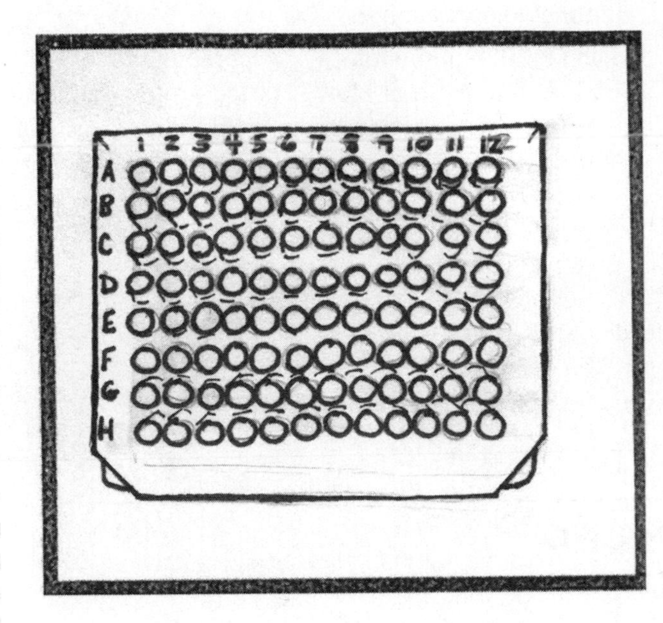

Objectives

- Load and develop an ELISA plate.
- Analyze and interpret your results.

DESCRIPTION

The Enzyme-Linked Immunosorbent Assay, or ELISA, is a method of testing for the presence of a chemical within a sample. This test uses antibodies to detect specific antigens, which are usually protein molecules. Antigens are molecules that are recognized by antibodies. Almost all proteins can be recognized as antigens, and some complex macromolecules can also be antigenic (e.g., the metabolites of certain illegal drugs, or the drugs themselves). Antibodies are created in laboratory animals and purified for use in ELISA testing. Typically, the antigen is purified from a sample (blood, tissue, etc.) and bound to the specially-coated plastic in a 96-well microtiter plate. Then, a blocking agent is added to the well. Blocking agents are proteins that will coat the plastic in the well that is not coated with antigen. The antibodies will not bind to the blocking agent. The primary antibody is then added, and will bind specifically to the antigen. A secondary antibody is then added, which will bind to the primary antibody. Secondary antibodies are conjugated with an enzyme that produces a colored product or light, after the addition of substrate. This color change or light emission is used to detect the presence of the antigen. The amount of antigen can be determined using a plate counter, a special machine designed to detect differences in color or light. ELISA tests may be qualitative or quantitative.

CONCEPTS & VOCABULARY

- Antigen
- B-lymphocyte (B-cell)
- Blocking
- Dot Blot Immunoassay
- Epitope
- Immunoglobin
- Primary and secondary antibodies
- Western blot

CURRENT APPLICATIONS

- There are many different types of analytical tests that use antibodies to detect the presence of an antigen. The ELISA test is a plate-based assay that allows for measurement of dozens of samples in a single preparation.
- The Western blot is a gel-based assay, where the sample is first separated on an SDS-PAGE gel, and then transferred to a nitrocellulose membrane for immunological analysis.
- The Dot Blot immunoassay is a combination of the ELISA and Western blot. A crude protein sample is spotted onto a membrane and analyzed using antibodies.

REFERENCES

Karp, G. (2010). The adaptive immune response; Uses of antibodies. *Cell and Molecular Biology: Concepts and Experiments*. 686–699; 763–765. Hoboken, NJ: John Wiley & Sons, Inc.

Voller, A, Bartlett, A, and Bidwell, DE. Enzyme immunoassays with special reference to ELISA techniques. *J. Clin. Pathol.* 31(6):507–520.

BACKGROUND

ANTIBODIES AND THE HUMORAL IMMUNE RESPONSE

Antibodies are large protein complexes produced by B-lymphocytes in mammals and birds. Antibodies are created by these cells to recognize foreign molecules, or *antigens*, in the bloodstream. Antibodies are composed of four peptides: two heavy chain and two light chain proteins. These four peptides form an immunoglobin complex: a Y-shaped protein with an antigen-binding region at the end of each of the two arms of the "Y." Each B-lymphocyte only produces one kind of antibody. A single antigen may have a number of sites recognizable by antibodies. These regions of recognition are called epitopes. The antibodies produced by a B-cell will only recognize one epitope. When an antigen is in the blood stream, it will be absorbed and degraded by other immune cells. After degradation, these cells will present certain epitopes on their plasma membrane. B-cells will initially present their specific antibodies on their plasma membrane. When the immunoglobin from a B-cell binds to the presented antigen on another immune cell, this will activate the B-cell. Once the B-cell has been activated, it will begin rapidly dividing, producing more of its specific antibody and releasing those antibodies into the blood stream. The antibodies will then bind to antigens and aggregate. This aggregation prevents both further infections and free movement of virus, bacteria, and other infectious agents. If a host cell is presenting antigens on its plasma membrane due to infection, antibodies will bind to the host cell and

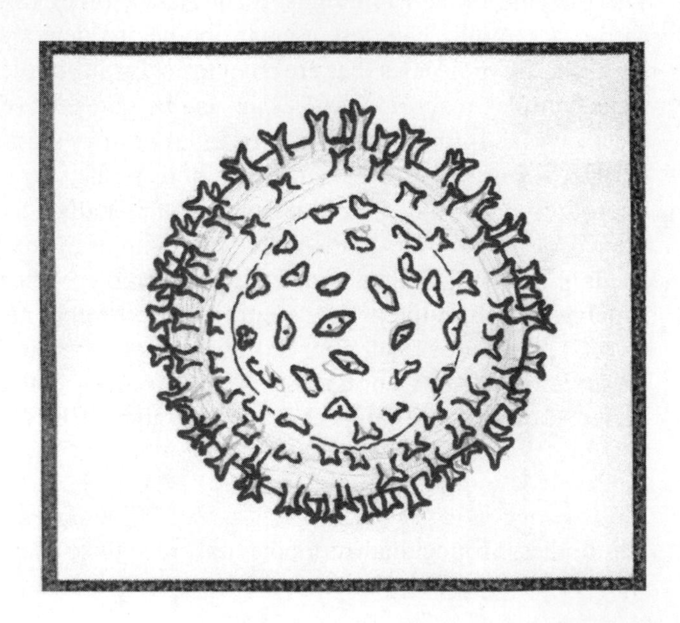

target it for destruction. After the infection has been completely cleaned out, the antibodies will be cleared from the blood stream, and the population of B-cells will decrease. However, there will always be more of the B-cells specific to that antigen in the blood stream. This will provide a more rapid secondary response if the organism is ever infected with the same infectious agent. This secondary immunity is what makes vaccinations effective. Vaccines are typically injections of non-infectious viruses, bacteria, or antigens associated with them.

USE OF ANTIBODIES IN BIOTECHNOLOGY

The molecular specificity of antibodies makes them a valuable tool in the laboratory. Antibodies can be produced to recognize most proteins. This is accomplished by injecting the antigen into a laboratory animal. These animals mount an immune response to the antigen, and then antibodies can be collected from their blood serum. This blood serum can be used as a polyclonal antibody. Polyclonal antibodies recognize multiple epitopes on a single antigen. This blood serum can be further purified to produce monoclonal antibodies. Monoclonal antibodies only recognize one specific epitope. Polyclonal antibodies can be used to recognize antigens in many molecular techniques, such as Western blots and the indirect ELISA test. Monoclonal antibodies are used when greater accuracy or specificity is required, such as in a sandwich ELISA. However, monoclonal antibodies are much more expensive to produce. Antibodies can also be used to detect antigens in microscopic samples (immunolocolization). To separate and purify antigens immunochromatography can be used, and may also be used to determine what other proteins bind to the antigens (immunoprecipitation or *pull-down*). Any application that requires the binding of an antigen can use antibodies in their design.

Antibodies that detect the antigens are called the primary antibodies. Primary antibodies can be conjugated to an enzyme or a fluorescent molecule for means of detection. Typically, this is not the case. A secondary antibody specific to an epitope on the immunoglobin of the organism used to create the primary antibody is usually employed. These secondary antibodies can be used for any primary antibody made by that organism. For example; a primary antibody specific to actin can be made in rabbit. It can then be detected with a secondary antibody specific to rabbit immunoglobin made in mouse. Secondary antibodies can be conjugated to a detection molecule. By using secondary antibodies, you remove both the error and cost involved with conjugating a detection molecule to your primary antibody. Also the secondary antibody can be used in any kind of antibodies produced in rabbit.

THE ELISA TEST

The ELISA test was designed to rapidly and accurately detect antigens of interest. Since its conception, the ELISA test has become a common tool in the medical field, law enforcement, the environmental sciences, and molecular biology. The simplicity of the application of the ELISA test makes it much more accessible than most of the other methods of protein detection. However, with simplicity comes added expense. The design and optimization of an ELISA test can be quite costly. However, once the ELISA test has been designed, so long as it has proper controls, its accuracy is near perfect.

The indirect ELISA test uses plastic 96-well microtiter plates that are specially coated to bind proteins. Other microtiter plates can be used, such as the 384-well plates. But the 96-well plates are the standard for laboratory use. Diluted, a purified antigen is added to a well. After an incubation period, the antigen is removed, and a blocking agent is added. The blocking agent is a protein that the

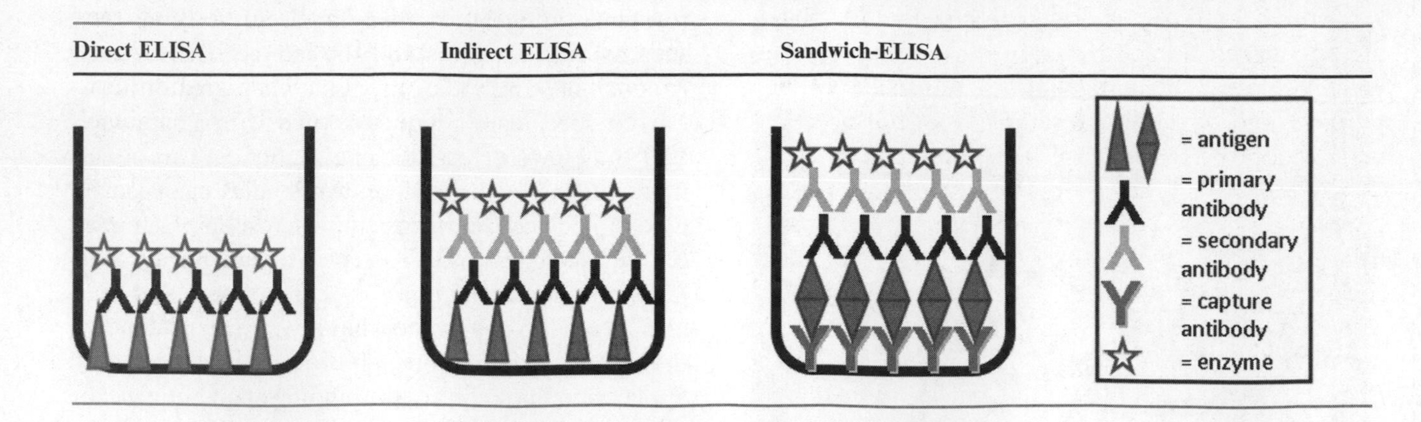

antibodies will not bind to. Bovine serum albumin or milk proteins are two commonly used blocking agents. After blocking, any spots in the well that have not bound the antigen will be coated with the blocking agent. This will prevent non-specific binding of the primary antibody to the walls of the well. The blocking agent is removed and the primary antibody solution is added. The primary antibody will bind to the antigen, but to nothing else in the well. The primary antibody solution is removed, then secondary antibody solution is added. The secondary antibody will bind to the primary antibody, but to nothing else in the well. The secondary antibody solution is removed, then a substrate solution is added. The substrate will be catalyzed by the enzyme conjugated to the secondary antibody. This reaction will typically produce either a colored product or light, indicating the presence of the antigen. This can be observed by eye, but many laboratories will use a plate reader to scan their ELISA. This not only allows for accurate detection, but also for a digital printout of the results.

The indirect ELISA test is used as a qualitative analysis tool. The direct ELISA only uses a primary antibody conjugated to an enzyme for detection. While this method is quicker and technically more accurate, it lacks the signal strength of the indirect ELISA. The sandwich ELISA test can be used to detect lower amounts of antigen and to obtain more quantitative results. The sandwich ELISA test requires at least one monoclonal antibody for the antigen of interest. It also requires that the antigen have two distinct epitopes that can be detected by antibodies. In the sandwich ELISA,

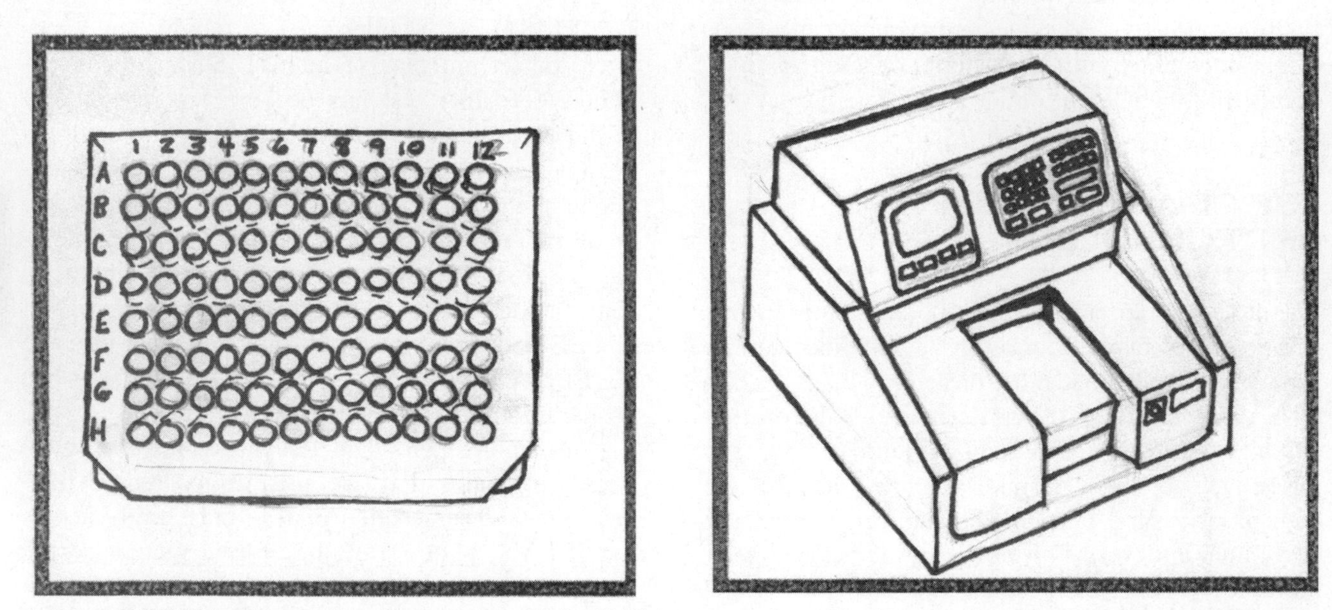

A 96-well ELISA plate and a plate-reader. This ELISA test used a colorimetric detection method and displays positive samples in column 2 and columns 6. Automated scanners can detect and measure the amount of color, allowing for quantitative measurements.

a captured antibody is added to the ELISA plate first. After blocking, the antigen is added to the ELISA plate. The antigen does not need to be purified, and typically the sample does not need to be diluted. The antigen will then bind to the captured antibody. Next a primary antibody is added, followed by a secondary antibody, a substrate solution and development and analysis of the ELISA plate. By using two antibodies specific to the antigen, sandwich ELISA allows for increased accuracy and decreased preparation for the test. However, the addition of a third antibody to the sandwich ELISA test will increase the cost. These types of tests are often used as a confirmation test to a positive result of a indirect ELISA or some other means of detection.

ELISA test kits are commonly used in the laboratory. A number of companies have produced ELISA tests for many different antigens. There are even a few ELISA test kits that are available for the general public. Familiarizing yourself with this test will serve you well as you pursue a career in science.

PROCEDURES

ELISA Protocol

1. Pretreated ELISA plates have been coated with diluted antigen incubated for 1 hour. The antigen solution was removed and blocking solution was added. The plates were and blocked for one hour before class at 37°C.
2. Dump the blocking solution (without washing or drying) and add primary antibody solution or 1% BSA/PBS, 25 μL per well, using the table below as a guide. You will perform a serial dilution on two controls and three experimental samples. Transfer 2.5 μL from Row A into Row B. Mix well, then transfer 2.5 μL from Row B into Row C. Repeat through row E. This should give you this dilution scheme: Row A: undiluted; Row B: 1 to 10; Row C: 1 to 100; Row D: 1 to 1,000; Row E: 1 to 10,000.
3. Seal with micropore tape and incubate at 37°C for 1 hour.
4. Dump the antibody solution and wash 5 times with 1× PBS/TWEEN-20.
5. Add 25 μL secondary antibody enzyme conjugate solution.
6. Seal with micropore tape and incubate at 37°C for 1 hour.
7. Dump secondary antibody enzyme conjugate solution and wash 5 times with 1 × PBS/TWEEN-20.
8. Add 25 μL substrate solution.
9. Seal with micropore tape, and develop at room temperature, reading by eye or with a microplate reader after 15 minutes, 30 minutes, and 1 hour.
10. Record your results on the next page. Please feel free to take a picture of your ELISA plate and add it to your worksheet.

ROW	Buffer Alone	Positive Control	Negative Control	Sample X	Sample Y	Sample Z
A	25 μL PBS	25 μL antibody	25 μL antibody	25 μL antibody	25 μL antibody	25 μL antibody
B	25 μL PBS	22.5 μL PBS	22.5 μL PBS	22.5 μL PBS	22.5 μL PBS	22.5 μL PBS
C	25 μL PBS	22.5 μL PBS	22.5 μL PBS	22.5 μL PBS	22.5 μL PBS	22.5 μL PBS
D	25 μL PBS	22.5 μL PBS	22.5 μL PBS	22.5 μL PBS	22.5 μL PBS	22.5 μL PBS
E	25 μL PBS	22.5 μL PBS	22.5 μL PBS	22.5 μL PBS	22.5 μL PBS	22.5 μL PBS
F	EMPTY	EMPTY	EMPTY	EMPTY	EMPTY	EMPTY
G	EMPTY	EMPTY	EMPTY	EMPTY	EMPTY	EMPTY
H	EMPTY	EMPTY	EMPTY	EMPTY	EMPTY	EMPTY

WORKSHEET

Observations and Assessments

1. Use the charts below, record your ELISA results. For the plate diagram, use the following symbols: 0 = no signal; + = weak signal; ++ = good signal; +++ = strong signal. You may also wish to shade in the circles, or paste a photograph of your plate onto this page for better visualization. If you have access to a microplate reader, record your quantitative data in the empty table below.

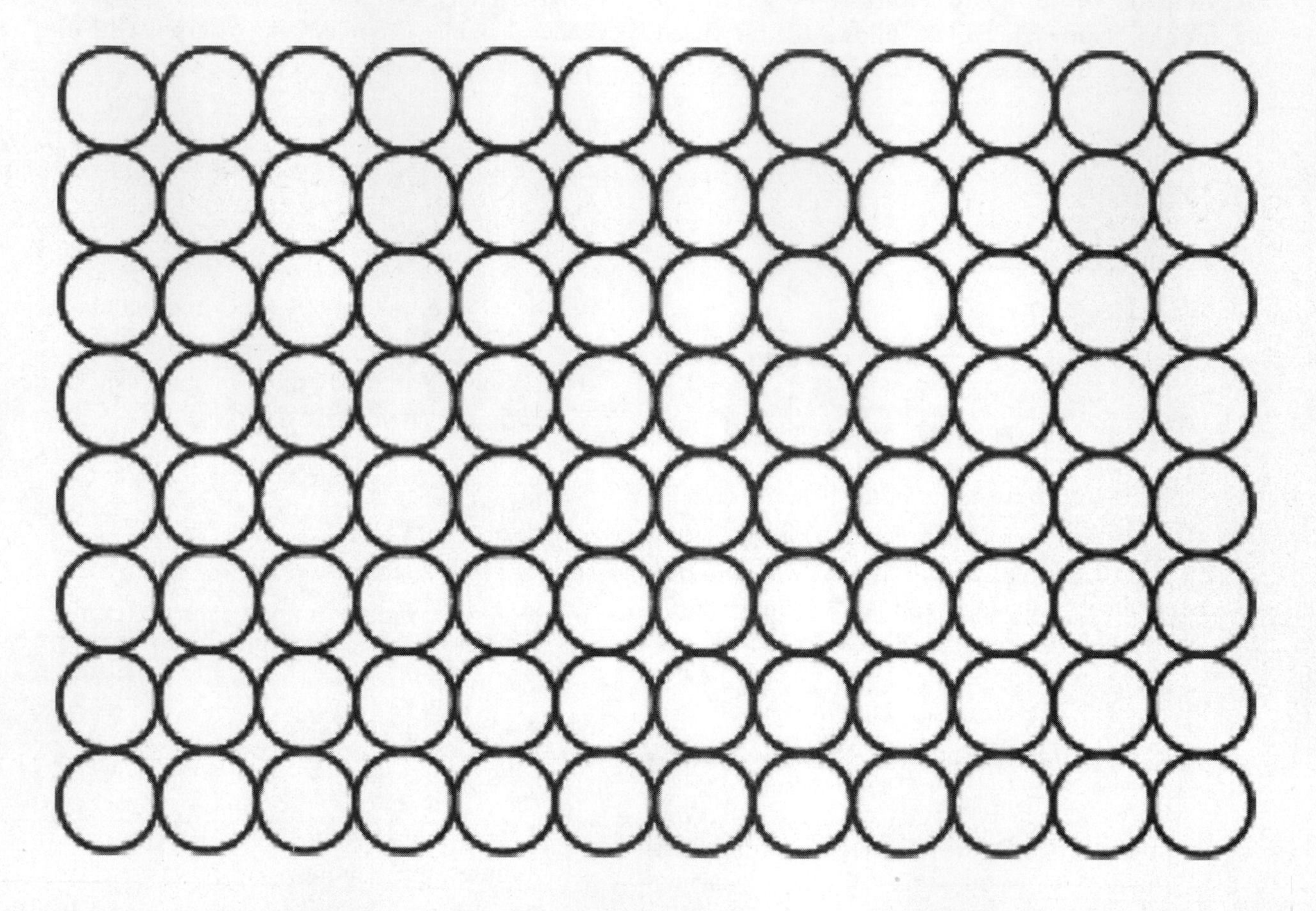

ROW	Buffer Alone	Positive Control	Negative Control	Sample X	Sample Y	Sample Z
A						
B						
C						
D						
E						
F						
G						
H						

2. Do your results match your expectations for the signal strength in the control samples? Did you see any signal in the positive control lanes? Does this signal strength decrease with each dilution in your series? Do you see any signal in the blank or negative control? What does deviation from your expectations mean for the accuracy of this ELISA test?

3. ELISA testing is commonly used to determine the presence of proteins in a substance. Many structural proteins are solid at room temperature and in aqueous solution, and must be solubilized using detergents or other chemicals before they can be used for liquid testing. The proteins found in solid silk fibers are typically hypoallergenic. Could an ELISA test be used to measure the amount of silk in the fabric of a shirt? Explain your answer.

4. What are the requirements for choosing a target for an ELISA test? What traits must the antigen of interest possess? What antigens would make poor targets for an ELISA? What antigens would make the best targets for ELISA testing?

5. You are a scientist working for the Drug Enforcement Agency. Your superiors have asked you to design an ELISA test to detect a new street drug in a suspect's blood sample. Describe how you would design the test. Focus on what antigen(s) you would target, what organisms you would use to produce the antibodies, and how you would optimize the test for general laboratory use.

DISCUSSION

Discussion Questions

1. Define antigen and epitope. How do they are relate to antibodies (immunoglobins)?
2. Discuss the possible uses for ELISA testing. Feel free to use library or Internet resources to expand your knowledge.
3. Why is the Blocking step necessary? What possible results would you observe if the blocking step were skipped?
4. What are the sources for error in an ELISA test? How has the test been designed to minimize this error?
5. Describe the differences in your results at each observation time. Discuss reasons for these differences, if you observe any.

Southern and Northern Blotting

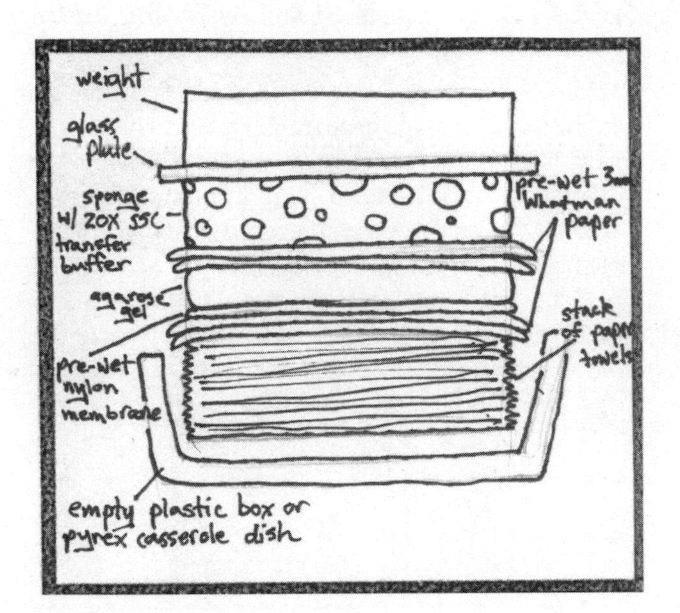

Objectives

- Transfer nucleic acids from an agarose gel to a nylon membrane.
- Practice the washing techniques involved with Southern/Northern Blot Analysis.
- Safely test your technique for handling radioactive isotope and X-ray film.

DESCRIPTION

Edwin Southern invented a technique in 1975 that allowed him to detect the presence or absence of specific sequences of genomic DNA. This method involved a restriction digest of the genomic DNA followed by agarose gel electrophoresis, and then transfer of the denatured DNA to a nitrocellulose membrane. Southern then used short oligonucleotide probes of a specific sequence to search for his DNA sequences of interest. A later technique was devised to search through RNA sequences also using oligonucleotide probes. This technique was called Northern blotting. Using Southern blotting and Northern blotting, one can determine the presence of genes or genetic sequences within a genome, or the presence of RNA transcripts being expressed by the cell. Today you are going to practice setting up the capillary transfer of an agarose gel and the blocking and hybridization of an oligonucleotide probe to a nylon membrane.

CONCEPTS & VOCABULARY

- Autoradiography
- Blocking agent
- cDNA library
- Capillary transfer
- Chemiluminescent
- Cross-linking
- Hybridization
- Nitrocellulose
- Oligonucleotide probe

CURRENT APPLICATIONS

- While there are many species of model organisms that can make use of a completely sequenced genome database, there are thousands more that cannot. Using a probe created from genetic sequence information from a known species, homologous genes may be discovered using Southern blot analysis. Then, using a DNA library and DNA sequencing technology, the identified genes can be sequenced and used for further research.
- While Northern blotting is still used to study the expression of a single gene transcript in an organism, new microarray technology makes studying the expression of the entire Transcriptome possible. A DNA microarray uses the same probe and complementary sequence hybridization as a Northern blot, but on a much larger scale. This technique is equivalent to hundreds of thousands of Northern blots, performed all at once. However, the technology is also about a thousand times more costly than a Northern blot. If you're only interested in a few genes, Northern blotting is the better method.

REFERENCES

Alwine, JC, et al. (1977). Method for detection of specific RNAs in agarose gels by transfer to diazobenzyloxymethyl-paper and hybridization with DNA probes. *Proc. Natl. Acad. Sci. USA.* 74(12): 5350–5354.

Karp, G. (2010). Nucleic acid hybridization. *Cell and Molecular Biology: Concepts and Experiments.* (745–746). Hoboken, NJ: John Wiley & Sons, Inc.

Southern, EM. (1975). Detection of specific sequences among DNA fragments separated by gel electrophoresis. *J. Mol. Biol.* 98(3): 503–517.

BACKGROUND

SOUTHERN AND NORTHERN BLOTTING

Edwin Southern invented a technique in 1975 that allowed him to detect the presence or absence of specific sequences of genomic DNA. This method involved a restriction digest of the genomic DNA followed by agarose gel electrophoresis separation of the restriction fragments. Southern then denatured the DNA by bathing the gel in sodium hydroxide. He then transferred the denatured DNA to a nitrocellulose membrane using capillary transfer as a means of translocation. Southern then used ultraviolet light to cross-link the denatured single-stranded DNA to the membrane and baked the membrane for a short time to fix and preserve his samples.

After he had this membrane with his digested genomic DNA sample adhered to it, he used short oligonucleotide probes of a specific sequence to search for his DNA sequences of interest. These tiny pieces of single-stranded DNA were labeled with radioactive phosphorus and later detected using X-ray film. The radiation from the probe would expose the X-ray film and then after development, the hybridized restriction bands on the membrane could be detected. This technique was named after Southern and today is called Southern blotting. Southern blotting is still used to search through

DNA samples for regions of known sequence. Most of the time genomic DNA is used, but even groups of PCR fragments, cDNA libraries, and other DNA samples may be used so long as they can be digested into smaller fragments and run on an agarose gel.

The probes of interest do not have to be created using DNA sequences from the organism that the scientist is working with. DNA sequences can be taken from other species and then used to search for similar gene sequences in closely related species.

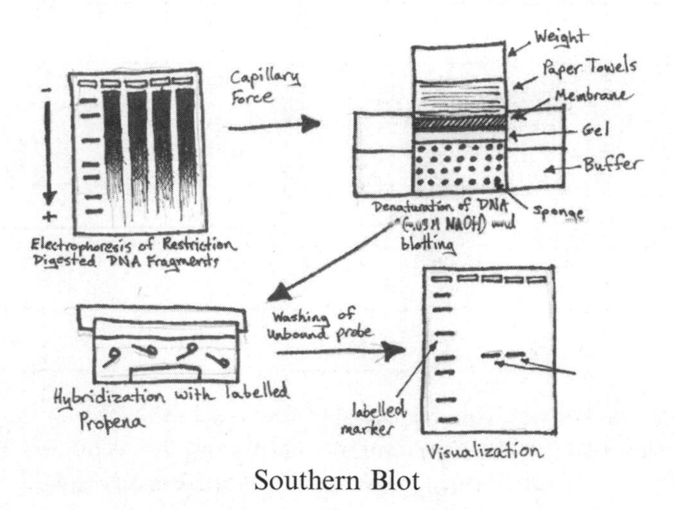

Southern Blot

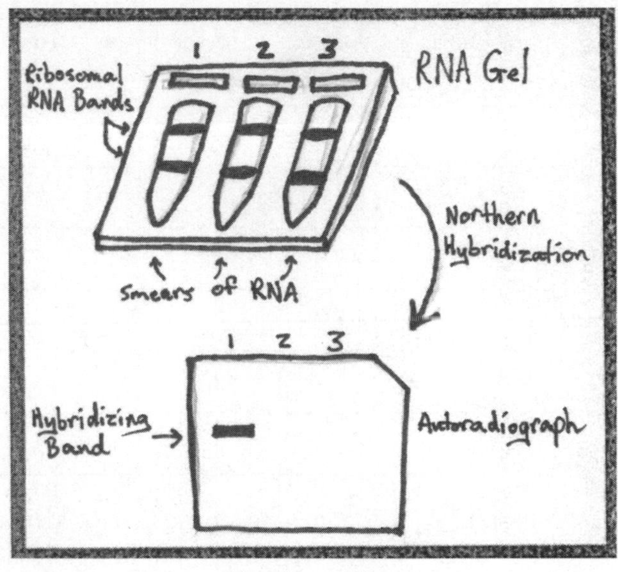

Northern Blot

PROCEDURES

Setting up the Capillary Transfer of an Agarose Gel to Nylon Membrane

1. Wash the gel with 0.25M HCl for 10 minutes at room temperature.
2. Rinse with distilled water.
3. Denature the gel in 0.5N NaOH/1.5M NaCl for 30 minutes at room temperature.
4. Neutralize the gel in 0.5M Tris (pH 7.25)/1.5M NaCl for 30 minutes.
5. Rinse with distilled water.
6. Set up the transfer using 10 × SSC Transfer Buffer, following the diagramed instructions.
7. Transfer overnight at room temperature.

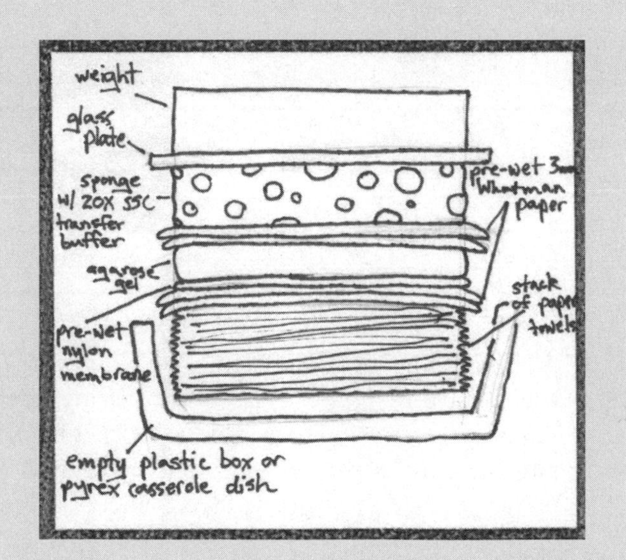

Blocking and Hybridization of a Nylon Membrane

1. Disassemble your capillary transfer setup, and remove the membrane. Wash the membrane with 2 × SSC Wash Buffer for 15 minutes at room temperature.
2. Cross-link the membrane under UV light for 2 minutes. Do not allow the membrane to dry out.
3. Pre-hybridize the blot for 15 minutes in hybridization buffer at 42°C.
4. Add the probe.
5. Hybridize at 42°C for 1 hour.
6. Wash the blot twice with Low Stringency Buffer for 5 minutes at 42°C.
7. Wash the blot once with High Stringency Buffer for 15 minutes at 42°C.
8. Wrap the blot in plastic wrap and set aside.
9. Using a UV-light, scan your person and workspace for fluorescence. This fluorescent tracking dye will tell you if you spilled any probe.

WORKSHEET

1. Paste UV-transilluminated photographs or drawings of your gel and post-transfer membrane below. Do the patterns match? If the patterns do not match, discuss the various possibilities why not. Compare the UV-fluorescence of the gel to the membrane. Estimate the amount of DNA transferred to the membrane, using the UV-fluorescence in the gel as a 100% initial concentration.

2. The nucleic acids in a gel must first be denatured before they can be transferred. Why must they be denatured? Discuss the differences in the denaturation of DNA samples versus RNA samples. Why are different methods employed for each type of molecule? Is the denaturing step the only difference between Southern and Northern blotting?

3. Discuss the difference between a Low-Stringency and a High-Stringency buffer. Why is the Low-Stringency buffer used to wash the membrane first? What are you washing away? Why is the High-Stringency buffer used to wash the membrane last? What are you washing away? Aside from changes to the buffers, what is another method that may be used to increase the stringency of a wash?

4. During your sweep of your person and workspace, did you find any fluorescent dye? If not, where did you find the dye? Discuss possible ways in which you accidentally contaminated your person or area. Discuss possible methods that could be employed in the future to reduce the possibility of contamination.

DISCUSSION

1. Genomic DNA must first be digested by restriction enzymes before it can be used for Southern Blotting. Why? What results would you expect if the restriction digest were skipped? Discuss the possible results from using different combinations of restriction enzymes.
2. Nucleic acids must be denatured prior to transfer to a membrane. DNA is typically denatured with sodium hydroxide or some other strong reducing agent. RNA is denatured with a mixture of formalin and formaldehyde. What is being denatured in the double-stranded DNA sample? What is being denatured in the single-stranded RNA sample? Discuss why this denaturation step is important.
3. You are a researcher working with goats. You are curious if goats have a gene within their genome that is similar to the PAX3 gene, which is linked to deafness and Waardenburg syndrome. Discuss an experimental design using the techniques that you learned in this experiment which would determine if a PAX3 homolog exists in goats.
4. You are a clinical medical researcher. You have identified a gene that may be involved in high blood cholesterol in Native American women. You want to know if this gene is increased or decreased in expression in Native American women with high blood cholesterol. Discuss an experimental design using the techniques that you learned in this experiment that would determine this information. Focus on your probe construction and your choice of samples. Discuss what a high expression level might mean when correlated to a high blood cholesterol concentration. Discuss what a low expression level might mean when correlated to a high blood cholesterol concentration.
5. Using the Internet, find one current example of a microarray used to test nucleic acid samples. Define what is being tested by the microarray, what samples are needed to use the technique, and what kind of data can be expected. Discuss research or experimentation that employed this technique.

Using Online Research Tools

Objectives

- To become familiar with online resources available for genetic studies.
- To research a human monogenic disorder.

DESCRIPTION

This exercise is designed to introduce you to the various online resources available to researchers via the Internet. Today we will focus on the free, public databases and search engines, but keep in mind that there are a number of private and/or subscription-based databases to which researchers may also have access. With the ease of information transfer, the ever-increasing power of computational algorithms, and the massive genome sequencing and proteomics efforts currently in progress, today is an incredible time to be a biologist. The full gene sequences of dozens of species are currently available, with hundreds more planned during the next decade. Researchers are also delving into the variations in transcript level/gene expression, and protein and metabolite levels under many different experimental conditions. All of this information is made available to researchers in the comfort of their home or office. Today, we will focus on human diseases caused by a single gene/protein mutation. Using this monogenic trait, you will learn about online tools in the design of a human gene therapy experiment.

CONCEPTS & VOCABULARY

- Familial
- Genomics
- Metabolomics
- Proteomics
- Sporadic
- Transcriptomics

CURRENT APPLICATIONS

- Almost every new experiment involving molecular cell biology also involves an online database search. As a researcher, there is no better skill than to be familiar with the online resources for a species that you are interested in. These websites are becoming more intuitive. The only problem that you may face is keeping current, as new content is added every year.

REFERENCES

Karp, G. (2010). Nature of the gene and genome; and the control of gene expression. *Cell and Molecular Biology: Concepts and Experiments.* (379–418; 419–422). Hoboken, NJ: John Wiley & Sons, Inc.

BACKGROUND

IMPORTANT WEBSITES

National Center for Biotechnology Information: For PubMed, Dan/RNA/protein Sequence & BLAST.
http://www.ncbi.nlm.nih.gov/

Online Mendelian Inheritance in Man: For information on human traits and genes.
http://www.ncbi.nlm.nih.gov/omim/

pdf file for Genes and Disease Book from NCBL: An excellent resource for people interested in Human Genomics, Gene Therapy, and Medical Genetics. It has hyperlinked human genome for easy searching.
http://www.ncbi.nlm.nih.gov/books/bookres.fcgi/gnd/pdf.html

Gene Therapy and the Human Genome Project: Some good information from the Oak Ridge National Laboratory and the Department of Energy.
http://www.ornlgov/sci/techresources/Human_Genome/medicine/medicine.shtml

Gene Tests from the NIH: One of my favorite resources for Molecular Medicine. Has a complete database of all available information on human genetic diseases (sporadic and familial) including prognosis, treatment information and physicians who specialize in each disease.
http://www.geneclinics.org/

HUMAN DISEASE RESEARCH EXERCISE

This laboratory exercise is designed to help you to become more aware and comfortable with the available online resources for biologists. This will be your guide into these websites and the appropriate use of their search engines.

To encourage you to make the most of this experience, your exercise will focus on a genetic disease.

BEFORE you come to class,

- You should choose a **monogenic**, hereditary disorder; or a genetic disease that is caused by a mutation in only one gene.
- This disease can be **familial** (inherited from a parent) or sporadic (caused by a random mutation), but hereditary disorders will be better studied (and easier to research).

Later, in class, you will become familiar with a variety of free, searchable online databases. Your instructor will guide you to determine a number of key facts about your chosen gene and its gene product. With that information in hand, you will design a gene therapy project to treat or potentially cure your chosen disease, and then read through the literature to find three papers that discuss current research about your chosen gene or disease.

PART 1: FINDING THE DISEASE

Complete **BEFORE** you come to class.

1. You must choose a genetic disorder to study for this exercise. It must be monogenic, and a hereditary disease will be easier to research. However, most genetic diseases are caused by random/sporadic mutations. If you wish to study a disease with a sporadic cause, be ready for a less straightforward endeavor. Also, be aware that some genetic diseases have both sporadic and familial varieties. Fill out the datasheet with your answers.
2. See if you can answer these questions: Is your chosen disease caused by a single gene? Are there other copies of this gene within the human genome? Within the human genome, there can exist several copies of a gene that are expressed in different cell types at different times. These isoforms can perform entirely different functions, but still code for the exact same gene. Can mutations in these isoforms cause similar disease symptoms? Remember, to keep your focus on only one disease. Fill out the datasheet with your answers.

PROCEDURES

To complete these two questions, start by reading through Chapters 19 and 20 in your Karp text. Then, using the websites provided on the **Overview** page of this chapter, begin to search for a disease that interests you. For the basic, preliminary information about a disease you should:

- Look through the "Genes and Disease" eBook from NCBI. This eBook has many common human diseases caused by some defect in a gene or gene product. Browse through this eBook, or search for a specific disease that may interest you.
- The "Gene Therapy and the Human Genome Project" website from the U.S. Department of Energy will give you background information about current human molecular medicine research and clinical trials.
- The Online Mendelian Inheritance in Man (OMIM) database is a no-frills resource that will give you all of the answers that you require for this project, but it is not very intuitive and may not be the most helpful resource for this project. Take a look at it, and see what you can glean from the entry on your disease.
- After you have chosen a disease, I highly recommend searching the Gene Reviews available at the Gene Tests website from the National Institutes of Health.
- I realize that many of you will be drawn to Wikipedia or WebMD to find out more information about your disease, and there is nothing wrong with that to start with. However, make sure that you verify all information with one of these other websites.

Part 2: Finding the Gene/Protein

Complete this work **IN CLASS**.

Your instructor will instruct you on how to use a number of different search engines available on the NCBI website. You will need to perform an Entrez Gene search in order to find information about your chosen gene, its transcript, and its gene product. You will use the NCBI Sequence Viewers to accomplish this. You will need to print out your search results, and highlight the information that you used to answer problem 3. Your instructor will also show you how to do an OMIM search and use PubMed to look for references about your gene/disease. If you keep your eyes open, you may also discover several important references as you perform your gene and protein searches, as well.

1. You will need to use the time with your instructor wisely and answer all of the questions for problem 3.
2. Fill out the datasheet with your answers and print out and attach your search results to the datasheet, making sure to highlight and label any important information.

Part 3: Understanding your Findings

Complete this work **DURING** class with your instructor.

After you have learned about the online resources and collected your data, you can tackle problem 4.

1. After you have completed problem 4, move on to problem 5 and look into the actual research projects that are being done to treat or cure your chosen disease.
   - Find three papers, and discuss the similarities and differences between their projects and your ideas from problem 4.
   - Does any of the information in these papers reveal any pitfalls in your idea?
   - Is your idea better than theirs?
   - Would you say that there is an abundance of research being performed in your chosen disease?
   - List your references in proper format.
   - Fill out the attached datasheet with your 3 references clipped to your answer sheet.

WORKSHEET

Write down your answers on a separate page of paper.

1. Which monogenic, hereditary disorder did you choose? Is it familial or sporadic? If it can be either both, which is more common?

2. Is it caused by a mutation in only a single gene? Do isoforms, or genes with similar sequence and function, exist for this gene? Can mutations in these other genes cause similar disease symptoms?

3. Answer the following questions:
   - What chromosome is your disease gene found on?
   - What is the map location for this gene (example, Xq27.3 for Fragile-X syndrome)?
   - How many nucleotides is this gene coding region?
   - How many nucleotides are in the gene's transcript?
   - How many amino acids are in this gene's peptide?
   - What is the function of this gene's peptide?
   - Where in the cell and body is this peptide typically found?

4. Using the information above, design a very basic gene therapy or molecular medical therapy that might be useful in treating the disease that you've chosen.
   - Describe your basic gene therapy method.
   - What is the specific mode of therapy (transformation, supplementation, gene silencing, etc.)?
   - Does the method use an *in vivo* vector?
     (vector choices, application of vector/RGT protein, etc.)
   - How will physicians apply this treatment?
     Does it involve cell transformation (via infection)?
     Injection of recombinant proteins?
   - What difficulties may arise during treatment?
     (difficult to get treatment to tissue, short half-life of activity, etc.)

5. Using the NCBI PubMed search engine, find three recent publications that discuss research/therapy involving your chosen disease gene. Are their therapy ideas similar to yours? Attach the first page of each reference to your assignment.

INDEX